MATH

&

Mathematicians

MATH & Mathematicians:

The History of Math Discoveries Around the World

Volume 4

Leonard C. Bruno
Lawrence W. Baker, Editor

Detroit • New York • San Diego • San Francisco • Cleveland • New Haven, Conn. • Waterville, Maine • London • Munich

Math and Mathematicians: The History of Math Discoveries Around the World, volume 4

Leonard C. Bruno

Project Editor
Lawrence W. Baker

Permissions
Kim Davis

Imaging and Multimedia
Leitha Etheridge-Sims, Robyn Young

Product Design
Cynthia Baldwin, Kate Scheible

Composition
Evi Seoud

Manufacturing
Rita Wimberley

For permission to use material from this product, submit your request via Web at http://www.gale-edit.com/permissions, or you may download our Permissions Request form and submit your request by fax or mail to:

Permissions Department
The Gale Group, Inc.
27500 Drake Rd.
Farmington Hills, MI 48331-3535
Permissions Hotline:
248-699-8006 or 800-877-4253; ext. 8006
Fax: 248-699-8074 or 800-762-4058

Permission for front cover biographical photographs are as follows (left to right): top row—Charlotte Angas Scott (Public domain), William Oughtred (Hulton/Archive by Getty Images), and John Nash (Corbis Corporation); middle row—Bertrand Russell (Library of Congress) and Stephen Hawking (AP/Wide World Photos); and bottom row—Karl Weierstrass (Corbis Corporation), Etta Zuber Falconer (Etta Zuber Falconer), and Edmond Halley (Library of Congress). Additional front cover permission: abacus (the Corbis Corporation) and U.S. Air Force Chapel (Colorado Springs Convention & Visitors Bureau). Permission for back cover photographs (left to right): Rock and Roll Hall of Fame (Convention & Visitors Bureau of Greater Cleveland), tree ring (the Stock Market), sundial (the Corbis Corporation), and dice (Field Mark Publications).

ISBN 0-7876-6481-2
Printed in the United States of America
10 9 8 7 6 5 4 3 2 1

Contents

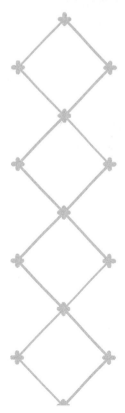

Contents

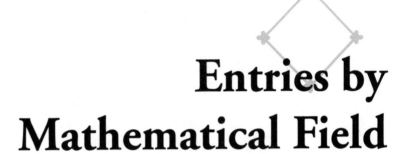

Entries by Mathematical Field

Algebra

Boldface type indicates volume number; regular type indicates page numbers.

Calculations

Calculus

Communications theory

Computer science

Group theory

Logic

Mathematical physics

Mathematics education

Mathematics in art

Mathematics scribe

Number theory

Probability and ratio

Pure mathematics

Set theory

Entries by Mathematical Field

Biographical Entries by Ethnicity

Boldface type indicates volume number; regular type indicates page numbers.

Biographical
Entries by
Ethnicity

Biographical
Entries by
Ethnicity

Reader's Guide

Mathematics has been described by one historian as "a vast adventure in ideas," but if that is so, we must always remind ourselves that it is individuals—real people—who have those ideas. However, unlike many other fields of science, mathematics seems to have only a few really well-known individuals whose names most people easily recognize. By high school, we all know something about the contributions of the mathematical greats like Euclid, Pythagoras, and Isaac Newton, yet the history of mathematics contains many more individuals whose accomplishments were nearly as important and whose lives may have been even more interesting. Volumes 1 and 2 of *Math and Mathematicians: The History of Math Discoveries Around the World* covered the early life, influences, and careers of fifty such individuals, telling the stories of those greats and near-greats whose contributions are not as well known as they should be. Volume 3 continued with thirty-two more mathematicians.

In volume 4 of *Math and Mathematicians,* thirty additional individuals have been selected on the same basis as the first three volumes: to describe and explain their mathematical contributions but also to offer readers a sampling of how rich the history of mathematics is and how diverse are its contributors. The thirty biographical entries in this volume include men and women from the ancient and modern world, who lived in nearly every major part of the globe. As with those individuals selected for the first three volumes, they too are linked by a common theme: that genius, hard work, determination, inspiration, and courage are multicultural, multiracial, and blind to gender differences.

Taken as a group, the thirty biographies in this volume include twelve people who were either born or did their major work in the twentieth century, five of whom are still living. This surely tells us that not all the great mathematicians are found only in old history books. The oldest historical figure included here, Greek geometer Apollonius of Perga, is thought to have thrived around 200 B.C.E., while the youngest living mathematician in this volume, American mathematician and educator John Allen Paulos, was born in 1945. Ranging over the entire history of mathematics and making selections for only thirty individuals for this volume obviously suggests that many more mathematicians were excluded than included, so it is not surprising that some truly deserving individuals were left out. However, with the help of an advisory board made up mostly of middle school and junior high librarians, we selected a solid group of mathematicians that spans the centuries.

In addition to the mathematical accomplishments that earned these individuals a high place in the history of mathematics, the varied and fascinating mix of tales make them extremely interesting from a human perspective. As might be expected, there were the usual child prodigies, like Indian mathematician Bhāskara II and French number theorist Émile Borel. Through an almost accidental set of circumstances, American computer scientist John Backus found that he possessed a love of mathematics that he then channeled toward the creation of a revolutionary computer language. Talented Canadian algebraist Agnes Sime Baxter abandoned a mathematical career in order to support her husband's career and to raise a family. As a high school senior, German American algebraic geometer Richard Courant found himself earning so much money as a tutor, that he had to stop going to class. One of the more bizarre legends of mathematics was German calculating prodigy Zacharias Dase who is today recognized to have been an "idiot savant." Called the "lightning calculator" because of his extraordinary ability to calculate in his head, Dase could not translate his dazzling calculating ability to any other part of his life and was at best average and sometimes even considered slow. Others were by no means so restricted, as Frenchman Girard Desargues was a talented architect as well as a geometer and engineer; M. C. Escher became the best-known graphic artist of his century; and Omar Khayyám became known more for his poems than for his mathematics. This parallels English mathematician and astronomer Edmond Halley, whose name is forever linked to his famous comet. Others had to overcome great odds, as Stephen Hawking did with Lou Gehrig's disease, John Nash did with schizophrenia, and Charlotte Angas Scott did by simply being a female.

The individuals in this volume bridge a span of over two thousand years and show us that progress in mathematics, as with any science, begins with a single enquiring mind simply wanting to understand a little bit more about a subject.

Added features

Math and Mathematicians: The History of Math Discoveries Around the World includes a number of additional features that help to make the connection between math concepts and theories, the people who discovered and worked with them, and common uses of mathematics.

- Three tables of contents, alphabetically by mathematician, by mathematical field, and by mathematician's ethnicity, provide varied access to the entries. Entries from volumes 1, 2, and 3 are included in the field and ethnicity tables of contents.

- A timeline provides a chronology of highlights in the history of mathematics.

- Dozens of photographs and illustrations bring to life the mathematicians, concepts, and ways in which mathematics is commonly used.

- Sidebars provide fascinating supplemental information about important mathematicians and theories.

- Extensive cross references make it easy to refer to other mathematicians and concepts covered in volumes 1 through 4; cross references to other entries are boldfaced upon the first mention in an entry.

- Sources for more information are found at the end of each entry so students know where to delve even deeper.

- A comprehensive cumulative index quickly points readers to the mathematicians, concepts, theories, and organizations mentioned in *Math and Mathematicians*. Entries from volumes 1, 2, and 3 are included in the index.

Special Thanks

The author wishes to thank his wife, Jane, and his three children, Nat, Ben, and Nina, for again giving me their patience and understanding. None ever made me feel that this book was a rival or unwanted competitor for my time. In fact, all actually helped at one time or another, usually when I needed to understand something or have it explained to me in a simple, direct way.

Thanks also go to freelancers Mya Nelson, for her detailed copyediting; Leslie Joseph, for her eagle-eyed proofreading; and Theresa Murray, for her concise indexing. Much appreciation also goes to Marco Di Vita at the Graphix Group for his fine typesetting work.

At U•X•L, I must first thank senior market analyst Meggin Condino for again offering me the opportunity to work on another fine U•X•L project. My deepest debt, however, goes to senior editor Larry Baker who has the talent and humanity to be whatever the situation calls for. One of the pleasures of doing this book was renewing my long-distance friendship with this paragon of hard work and high standards. His sense of humor, total mastery of his craft, and easy understanding of a writer's situation make him able to handle virtually anything a writer can throw at him. Finally, he is not averse to simple hard work. I only wish we had more time to talk baseball.

Comments and suggestions

We welcome your comments on *Math and Mathematicians* as well as your suggestions for biographies to be featured in future volumes. Please write: Editors, *Math and Mathematicians*, U•X•L, 27500 Drake Rd., Farmington Hills, Michigan, 48331-3535; call toll-free: 1-800-347-4253; fax to 248-699-8097; or send e-mail via www.gale.com.

Advisory Board

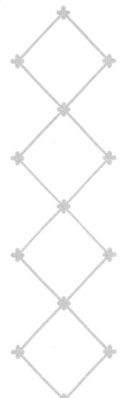

Kari Deck
Librarian, Jim Hill Middle School, Minot, North Dakota

Jacquelyn Divers
Librarian, Cave Spring Middle School, Roanoke, Virginia

Elaine Ezell
Library Media Specialist, Bowling Green Junior High School, Bowling Green, Ohio

Marie-Claire Kelin
Library Media Services Teacher, Lincoln Middle School, Santa Monica, California

Eric Stromberg
Assistant Principal/Former Mathematics Teacher, Riley Middle School, Livonia, Michigan

Milestones in the History of Mathematics

50,000 B.C.E.   Primitive humans leave behind evidence of their ability to count. Paleolithic people in central Europe make notches on animal bones to tally.

15,000 B.C.E.   Cave dwellers in the Middle East make notches on bones to keep count and possibly to track the lunar cycle.

c. 8000 B.C.E.   Clay tokens are used in Mesopotamia to record numbers of animals. This eventually develops into the first system of numeration.

3500 B.C.E.   The Egyptian number system reaches the point where they now can record numbers as large as necessary by introducing new symbols.

75,000 B.C.E.
Neanderthal man
can communicate
by speech.

38,000 B.C.E.
Homo sapiens
species evolves from
Neanderthal man.

12,000 B.C.E.
The dog is
domesticated from
the Asian wolf.

8000 B.C.E.
Earth's human
population soars to
5.3 million.

50,000 B.C.E.   40,000 B.C.E.   30,000 B.C.E.   20,000 B.C.E.   8000 B.C.E.

Milestones in the History of Mathematics

c. 2400 B.C.E.	Mathematical tablets dated to this period are found at Ur, a city of ancient Sumer (present-day Iraq).
c. 2000 B.C.E.	Babylonians and Egyptians use fractions as a way to help them tell time and measure angles.
c. 1800 B.C.E.	The Babylonians know and use what is later called the Pythagorean theorem, but they do not yet have a proof for it.
c. 1650 B.C.E.	The Rhind papyrus (also known as the Ahmes papyrus) is prepared by Egyptian scribe Ahmes, which contains solutions to simple equations. It becomes a primary source of knowledge about early Egyptian mathematics, describing their methods of multiplication, division, and algebra.
876 B.C.E.	The first known reference to the usage of the symbol for zero is made in India.
c. 585 B.C.E.	Greek geometer and philosopher Thales of Miletus converts Egyptian geometry into an abstract study. He removes mathematics from a sole consideration of practical problems and proves mathematical statements by a series of logical arguments. Doing this, Thales invents deductive mathematics.
c. 500 B.C.E.	Greek geometer and philosopher Pythagoras of Samos formulates the idea that the entire universe rests on numbers and their relationships. He deduces that the square of the length of the hypotenuse of a right triangle is equal to the sum of the squares of the lengths of its sides. It becomes known as the Pythagorean theorem.

3500 B.C.E.
Human civilization begins as the Sumerian society emerges.

3000 B.C.E.
The Sahara Desert has its beginnings in North Africa.

2485 B.C.E.
The Great Sphinx carved from rock at Giza.

776 B.C.E.
First recorded Olympic games in Greece are held.

625 B.C.E.
Metal coins are introduced in Greece.

3500 B.C.E.　　2750 B.C.E.　　2000 B.C.E.　　1250 B.C.E.　　500 B.C.E.

c. 440 B.C.E.	Greek geometer Hippocrates of Chios writes *Elements of Geometry,* regarded by many as the first mathematical textbook.

c. 300 B.C.E.   Greek geometer Euclid of Alexandria writes a textbook on geometry called the *Elements.* It becomes the standard work on its subject for over 2,000 years.

c. 240 B.C.E.   Greek geometer Archimedes of Syracuse calculates the most accurate arithmetical value for pi (π) to date. He also uses a system for expressing large numbers that uses an exponential-like method. Archimedes also finds areas and volumes of special curved surfaces and solids.

c. 230 B.C.E.   Greek astronomer Eratosthenes develops a system for determining prime numbers that becomes known as the "sieve of Eratosthenes."

225 B.C.E.   Greek geometer Apollonius of Perga introduces the terms "parabola," "ellipse," and "hyperbola" in his major eight-book work *The Conics.*

c. 100 B.C.E.   Negative numbers are used in China.

60 C.E.   Greek geometer Hero of Alexandria writes his most famous mathematical work, *Metrica,* a three-volume book on how to calculate and divide area and volume. In it, he defines geometry as "the science of measuring land."

c. 150   Greek geometer and astronomer Claudius Ptolemy's geometrical theories have important applications in astronomy.

429 B.C.E.
The plague kills at least one-third of the population of Athens, Greece.

214 B.C.E.
Construction begins on the Great Wall of China.

153 B.C.E.
January 1 becomes the first day of the civil year in Rome.

C. 6 B.C.E.
Jesus Christ is born.

500 B.C.E.   300 B.C.E.   100 B.C.E.   100 C.E.   200

c. 250   Greek algebraist Diophantus of Alexandria is the first Greek to write a significant work on algebra.

c. 320   Greek geometer Pappus of Alexandria summarizes in a book all acquired knowledge of Greek mathematics, making it the best source for Greek mathematics. French number theorist Pierre de Fermat later restores and studies Pappus's work.

c. 400   Greek geometer, astronomer, and philosopher Hypatia of Alexandria writes commentaries on Greek mathematicians Apollonius of Perga and Diophantus of Alexandria. She is the only woman scholar of ancient times and the first woman mentioned in the history of mathematics.

499   Hindu mathematician and astronomer Aryabhata the Elder describes the Indian numerical system. He also uses division to popularize a method for finding the greatest common divisor of two numbers.

700   Negative numbers are introduced by the Hindus to represent a negative balance.

820   Arab algebraist and astronomer al-Khwārizmī writes a mathematics book that introduces the Arabic word *al-jabr,* which becomes transliterated as algebra. His own name is distorted by translation into "algorism," which comes to mean the art of calculating or arithmetic. Al-Khwārizmī also uses Hindu numerals, including zero, and when his work is translated into Latin and published in the West, those numerals are called "Arabic numerals."

222 C.E
Chinese alchemists
invent gunpowder.

629
The Koran is
established as the
holy book of Islam.

752
Japan's 55-foot Buddha
statue is completed.

200   350   500   650   800

c. 825	Arab algebraist and astronomer al-Khwārizmī recommends the use of a decimal system.

1072	Persian mathematician, astronomer, and poet Omar Khayyám writes *Treatise on Demonstration of Problems of Algebra,* a landmark work that largely deals with cubic equations. He also measures the length of the year to be 365.24219858156 days, an amazingly accurate result.
c. 1140	Indian mathematician Bhāskara II writes a work on arithmetic and geometry called *Līlāvatī* (The Beautiful), which first uses decimal notation. He also writes about algebra in *Bījaganita* (Seed Arithmetic), in which he uses letters to represent unknown quantities.
1202	Italian number theorist Leonardo Pisano Fibonacci writes about the abacus, the use of zero and Arabic (Hindu) numerals, the importance of positional notations, and the merits of the decimal system.
1225	Italian number theorist Leonardo Pisano Fibonacci writes *Liber Quadratorum* in which he uses algebra based on the Arabic system.
1299	A law is passed in Florence, Italy, forbidding the use of Hindu-Arabic numbers by bankers. Authorities believe such numbers are more easily forged than Roman numerals.
1364	French algebraist and geometer Nicole d'Oresme writes *Latitudes of Forms,* an important work on coordinate systems.

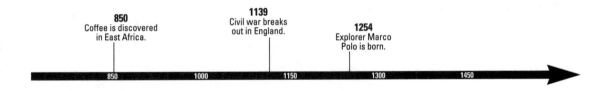

850
Coffee is discovered in East Africa.

1139
Civil war breaks out in England.

1254
Explorer Marco Polo is born.

850   1000   1150   1300   1450

c. 1470	French arithmetician and algebraist Nicolas Chuquet writes *Triparty en la science des nombres* (Three Parts in the Science of Numbers), the earliest French algebra book.
1482	The first printed edition of Greek geometer Euclid of Alexandria's geometry book, *Elements,* is published in Venice, Italy.
1489	The plus (+) and minus (–) symbols are first used in a book by German mathematician Johannes Widmann. They are not used as symbols of operation but merely to indicate excess and deficiency.
1535	Italian mathematician Niccolò Tartaglia demonstrates in a public forum his correct solution to the cubic equation. He later discloses in confidence his secret methods to another Italian mathematician, Girolamo Cardano, who later publishes the solution after learning that another Italian mathematician, Scipione dal Ferro, had discovered the solution as early as 1515. Cardano's paper correctly gives credit to both Tartaglia and dal Ferro.
1557	Welsh-born English mathematician Robert Recorde is the first to use the modern symbol for equality (=) in a book.
1570	The first complete English translation of *Elements,* by Greek geometer Euclid of Alexandria, appears.
1581	Italian mathematician Galileo discovers that the amount of time for a pendulum to swing back and forth is the same, regardless of the size of

1492
Christopher Columbus
discovers America.

1559
Mary, Queen of Scots,
becomes queen of England.

1585
Walter Raleigh
founds first colony
in Virginia.

1475   1500   1525   1550   1585

the arc. Dutch astronomer and mathematical physicist Christiaan Huygens would later use this principle to build the first pendulum clock.

1584   German algebraist Joost Bürgi begins work on improving a system of computing called "prosthaphairesis," a method of doing complicated multiplication by simple addition. By the end of the decade, he happened upon the idea of logarithms, and he created some actual conversion tables.

1585   Dutch mathematician Simon Stevin writes about the first comprehensive system of decimal fractions and their practical applications.

1591   French algebraist François Viète introduces the first systematic use of symbolic algebraic notation. He demonstrates the value of symbols by using the plus and minus signs for operations, vowels for unknown quantities (called variables), and consonants for known quantities (called parameters).

1594   Scottish mathematician John Napier first conceives of the notion of obtaining exponential expressions for various numbers, and begins work on the complicated formulas for what he eventually calls logarithms.

1601   English mathematician Thomas Harriot discovers the law of refraction of light.

1609   German astronomer and mathematician Johannes Kepler advances the development of the geometry of the ellipse as he attempts to prove that planets move in elliptical orbits.

1587
The first child of English parents is born in North America.

1590
William Shakespeare begins writing plays.

1603
Russian famine kills tens of thousands.

1585   1590   1595   1600   1610

1609	Italian mathematician Galileo improves upon the invention of the telescope by building a version with a magnification of about thirty times.
1614	Scottish mathematician John Napier invents "Napier's bones." This calculating machine consists of sticks with a multiplication table on the face of each stick. Calculations can be done by turning the rods by hand. He also publishes a book on logarithms.
1616	English mathematician Henry Briggs works with the Scottish inventor of logarithms, John Napier, to improve the base system of logarithms. Both agree that a base of 10 is the best method.
1619	German astronomer and mathematician Johannes Kepler shows that a planet's revolution is proportional to the cube of its average distance from the Sun.
c. 1621	English mathemetician William Oughtred invents the straight logarithmic slide rule.
1629	French number theorist Pierre de Fermat pioneers the application of algebra to geometry. Although French algebraist and philosopher René Descartes is credited with the invention and full development of analytic geometry, Fermat develops it earlier but does not publish his findings.
1631	English mathematician William Oughtred includes a large amount of mathematical symbolism in a book he publishes, including the notation "×" for multiplication and "::" for proportion.

1620
Pilgrims land in
Plymouth Colony.

1628
Salem, Massachusetts,
is founded.

1630
Lemonade is invented
in Paris, France.

1610   1615   1620   1625   1630

| 1632 | Italian mathematician Galileo discounts the theory of an Earth-centered universe. As a result, the Roman Inquisition sentences him to life imprisonment. |

| 1636 | French number theorist Pierre de Fermat introduces the modern theory of numbers. His work includes theories on prime numbers. |

| 1637 | French algebraist and philosopher René Descartes introduces analytic geometry by demonstrating how geometric forms may be systematically studied by analytic or algebraical means. He is the first person to use the letters near the beginning of the alphabet for constants and those near the end for variables. He also includes a notation system for expressing exponents. |

| c. 1637 | French number theorist Pierre de Fermat writes in the margin of a book a reference to what comes to be known as "Fermat's last theorem." This theorem remains the most famous unsolved problem in mathematics until it is solved in 1993 by Andrew J. Wiles. Fermat says he has a proof for the particular problem posed, but that the margin is too small to include it there. |

| 1639 | French geometer and engineer Girard Desargues begins the study of projective geometry. |

| 1642 | French geometer Blaise Pascal invents the first automatic calculator. It performs addition and subtraction by means of a set of wheels linked together by gears. |

| 1644 | French number theorist Marin Mersenne suggests a formula that will yield prime numbers. |

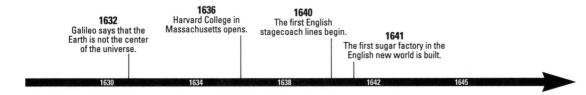

1632 Galileo says that the Earth is not the center of the universe.

1636 Harvard College in Massachusetts opens.

1640 The first English stagecoach lines begin.

1641 The first sugar factory in the English new world is built.

1630   1634   1638   1642   1645

These "Mersenne numbers" are not always correct, but they stimulate research into the theory of numbers.

1651 Danish mathematical astronomer Nicolaus Mercator publishes three books: *Trigonometria sphaericorum logarithmica,* which explains how to use logarithms in trigonometry; *Cosmographia,* which explains the physical geography of the Earth; and *Astronomia,* which gives a mathematical introduction to astronomy.

1654 French number theorist Pierre de Fermat exchanges letters with French geometer Blaise Pascal in which they discuss the basic laws of probability and essentially found the theory of probability.

1657 Dutch astronomer and mathematical physicist Christiaan Huygens writes about probability.

1659 Swiss mathematician Johann Heinrich Rahn is the first to use today's division sign ($\div$) in a book. Later, English mathematician John Wallis adopts it and popularizes it through his works.

1660 English geometer Isaac Barrow begins his professorship at Cambridge University. Nine years later, he begins a professional relationship with English physicist Isaac Newton.

1662 English geometer William Brouncker becomes the first president of the prestigious Royal Society of London, a position he would hold until 1677.

1662 English statistician John Graunt is the first to apply mathematics to the integration of vital sta-

1651
Leonardo da Vinci's
Treatise on Painting
is published.

1652
Capetown, South
Africa, is founded.

1659
Typhoid fever is
described for the
first time.

| 1650 | 1653 | 1656 | 1659 | 1662 |

tistics. As the first to establish life expectancy and to publish a table of demographic data, Graunt is considered the founder of vital statistics.

1666   German logician Gottfried Leibniz begins the study of symbolic logic by calling for a "calculus of reasoning" in mathematics.

1668   Danish mathematician and astronomer Nicolaus Mercator is the first to calculate the area under a curve using the newly developed analytical geometry.

1673   German logician Gottfried Leibniz begins his development of differential and integral calculus independently of English physicist Isaac Newton.

1674   Japanese mathematician Seki Kōwa publishes his only book, in which he solves 15 supposedly "unsolvable" problems.

1684   German logician Gottfried Leibniz publishes an account of his discovery of calculus. He discovers it independently of English physicist Isaac Newton, although later than him. Newton, however, publishes his discovery after Leibniz in 1687. The timing of the discovery produces a feud between the two men.

1687   English physicist Isaac Newton introduces the laws of motion and universal gravitation and his invention of calculus.

1690   Massachusetts is the first colony to produce paper currency.

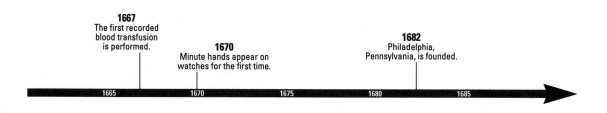

1667
The first recorded
blood transfusion
is performed.

1670
Minute hands appear on
watches for the first time.

1682
Philadelphia,
Pennsylvania, is founded.

1665   1670   1675   1680   1685

1693	English astronomer Edmond Halley compiles the first set of detailed mortality tables, making use of statistics in the study of life and death.
1706	English geometer William Jones is the first to use the sixteenth letter of the Greek alphabet, pi (π), as the symbol for the ratio of the circumference to the diameter of a circle.
1713	The first full-length treatment of the theory of probability appears in a work by Swiss mathematician Jakob Bernoulli.
1737	Swiss geometer and number theorist Leonhard Euler formally adopts the sixteenth letter of the Greek alphabet (π) as the symbol for the ratio of the circumference to the diameter of a circle. The ratio itself becomes known as pi. Following his adoption and use, it is generally accepted.
1742	Scottish geometer and physicist Colin Maclaurin writes *Treatise of Fluxions,* which logically and systematically explains the work of English physicist Isaac Newton. In this two-volume work, Maclaurin articulates Newton's calculus in great detail and lays out an excellent geometrical framework to support it.
1748	Italian mathematician Maria Agnesi publishes *Analytical Institutions,* a large, two-volume work that surveys elementary and advanced mathematics. Agnesi is best known for her consideration of the cubic curve or what comes to be translated as the "witch of Agnesi."
1750	Swiss geometer and probability theorist Gabriel Cramer discovers a simple procedure for solving

1704
America's first regular newspaper begins publication.

1705
Thomas Newcomen invents the steam engine.

1714
Daniel Fahrenheit builds a mercury thermometer.

1725
Antonio Vivaldi composes *The Four Seasons.*

1732
Benjamin Franklin revolutionizes the colonial postal service.

1690   1700   1710   1725   1740

systems of simultaneous linear equations. It becomes known as Cramer's Rule.

1755   Nineteen-year-old French algebraist Joseph-Louis Lagrange sends a paper to Swiss geometer and number theorist Leonhard Euler concerning Lagrange's "calculus of variations." Euler is so impressed with the young man's work that he holds back his own writings on the subject, thus allowing Lagrange priority of publication.

1758   English astronomer and mathematician Edmond Halley's predictions come true when a comet—which became known as Halley's comet—appears. Halley spent years searching historical records and studying the movements of two dozen other comets to arrive at his prediction.

1763   French geometer Gaspard Monge begins the study of descriptive geometry.

1767   Swiss-born German geometer Johann Lambert proves that the number for pi (π) is irrational.

1791   African American mathematician Benjamin Banneker assists in the surveying process of the new city of Washington, D.C.

1792   African American mathematician Benjamin Banneker publishes his first *Almanac*.

1792   The United States establishes its first monetary system, making the dollar its basic unit of currency.

1794   French geometer Gaspard Monge helps develop the metric system.

1795   France adopts the metric system.

Milestones in the History of Mathematics

1754 Seven Years' War between the French and Indians begins.

1776 Declaration of Independence is written.

1789 French Revolution begins.

1750   1760   1770   1780   1790

1797	German mathematician Carl Friedrich Gauss gives the first wholly satisfactory proof of the fundamental theorem of algebra.
1813	English mathematician Charles Babbage cofounds The Analytical Society, whose general purpose is to revive mathematical analysis in England.
1816	French mathematician Sophie Germain receives an award for her paper on the mathematical theory of elasticity.
1820	English mathematician Charles Babbage conceives of the idea of calculation "by machinery." Over the next fifty years, he works on developing the "difference engine," but never succeeds. The technical requirements for such a machine turn out to be beyond the engineering ability of his time.
1821	French mathematician Augustin-Louis Cauchy publishes the first of three books on calculus.
1825	Norwegian mathematician Niels Abel first proves the impossibility of solving the general quintic equation by means of radicals. This problem had puzzled mathematicians for two and a half centuries.
1829	Russian geometer Nicolay Lobachevsky describes his discovery of non-Euclidean geometry. This system includes the concept that an indefinite number of lines can be drawn in a plane parallel to a given line through a given point.

1794
Eli Whitney invents
the cotton gin.

1803
The United States
nearly doubles,
following the
Louisiana Purchase.

1818
Russian socialist leader
Karl Marx is born.

1827
Contact lenses
are invented.

1795 1800 1810 1820 1830

1829	German mathematical physicist Carl Jacobi publishes *Fundamenta nova theorae functionum ellipticarum,* which contains important work on the theory of elliptic functions.
1830	French algebraist and group theorist Évariste Galois is the first to use the word "group" in the technical sense and to apply groups of substitutions to the question of reducibility of algebraic equations.
1832	Hungarian geometer János Bolyai announces his discovery of non-Euclidean geometry, which he makes at about the same time as Russian geometer Nikolay Lobachevsky. His discovery is totally independent of Lobachevsky's, and when Bolyai finally sees Lobachevsky's work, he thinks it has been plagarized from his own.
1832	Swiss geometer Jakob Steiner lays the foundations for what would become known as projective geometry.
1833	Irish algebraist William Rowan Hamilton makes one of the first attempts at analyzing the basis of irrational numbers. His theory views both rational and irrational numbers as based on algebraic number couples.
1840	German calculating prodigy Zacharias Dase computes in his head pi to two hundred decimal places. His work is published in the prominent *Crelle's Journal* four years later.
1842	English applied mathematician Ada Lovelace translates an article about the computing machine ideas of English mathematician

Milestones in the History of Mathematics

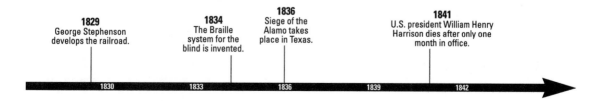

1829
George Stephenson develops the railroad.

1834
The Braille system for the blind is invented.

1836
Siege of the Alamo takes place in Texas.

1841
U.S. president William Henry Harrison dies after only one month in office.

| 1830 | 1833 | 1836 | 1839 | 1842 |

Charles Babbage. The notes she adds to the article produce the first clear mechanical explanation of Babbage's planned analytical engine and provide actual examples of how his machine might be instructed to perform certain tasks. These notes are now recognized as the world's first computer program.

1847 English logician George Boole maintains that the essential character of mathematics lies in its form rather than in its content. His work focuses on mathematics as symbolic rather than only "the science of measurement and number."

1854 English logician George Boole establishes both formal logic and Boolean algebra.

1854 German geometer Bernhard Riemann offers a global view of geometry. He develops further the ideas of Russian geometer Nikolay Lobachevsky and Hungarian geometer János Bolyai and introduces a new, non-Euclidean system of geometry.

1854 German analyst and number theorist Karl Weierstrass publishes his work on the mathematical functions named after Norwegian algebraist Niels Abel in the prestigious research journal *Crelle's Journal.*

1858 German number theorist Richard Dedekind conceives of the idea—later to be called the "Dedekind cut"—that treats the problem of irrational numbers in an entirely new manner, allowing irrational numbers to be categorized as fractions.

1844 Gottlob Keller invents the wood pulp paper process.

1846 Mexican War begins.

1856 Neanderthal man fossils are found.

1845   1848   1851   1854   1857

1860	German geometer Bernhard Riemann uses the complex number theory to form the basis for most of the research in prime numbers for the next century.
1874	German mathematician Georg Cantor begins his revolutionary work on set theory and the theory of the infinite and creates a whole new field of mathematical research.
1874	Russian mathematician Sofya Kovalevskaya writes two papers on differential equations.
1883	English algebraist and geometer Arthur Cayley becomes president of the British Association for the Advancement of Science.
1884	Greenwich, England, is chosen as the site where the world's 24 time zones begin.
1885	English analytical geometer and educator Charlotte Angas Scott receives her Ph.D. in mathematics from the University of London, the first woman in England to obtain a doctorate in mathematics.
1885	The newly opened Bryn Mawr College in Bryn Mawr, Pennsylvania, selects English analytical geometer and educator Charlotte Angas Scott to be the first head of its mathematics department.
1888	Russian mathematician Sofya Kovalevskaya receives an award for her paper on the problem of how Saturn's rings rotate the planet.
1891	Canadian algebraist Agnes Sime Baxter receives her bachelor's degree in mathematics with "first

Milestones in the History of Mathematics

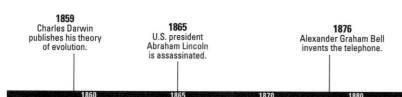

class distinction," the first female student at Dalhousie University given such an honor.

1891 English analytical geometer and educator Charlotte Angas Scott is the only woman on the first council of the New York Mathematical Society, which later evolved into the American Mathematical Society.

1896 French analyst Jacques-Salomon Hadamard is the first to offer a correct proof showing that there is an infinite number of prime numbers.

1903 English mathematical learning theorist Mary Everest Boole publishes *Lectures on the Logic of Arithmetic,* which, along with *The Preparation of the Child for Science,* published a year later, have a considerable influence on the progressive school movements of the early twentieth century.

1905 German American physicist and mathematician Albert Einstein writes five landmark papers that cover Brownian motion, the photoelectric effect, and his theory of relativity. It was with relativity that he devised his famous formula, $E = mc^2$.

1909 Danish mathematician Agner K. Erlang publishes "The Theory of Probabilities and Telephone Conversations." In this important work, he develops a formula that demonstrates the number of phone calls to arrive during a certain period of time, thereby allowing a phone company to calculate the fraction of callers who must wait when trying to place a call. This allows the company to provide more efficient service.

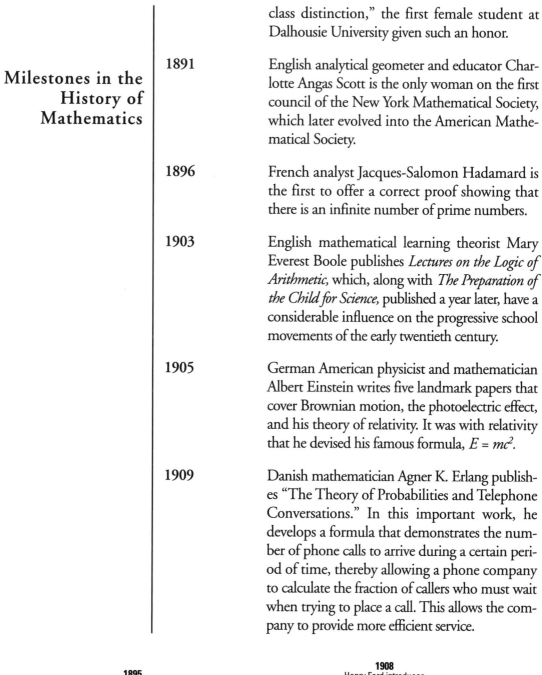

1895
Helium is
discovered.

1903
Wright Brothers take
first airplane flight.

1908
Henry Ford introduces
the Model T.

1892   1896   1900   1904   1910

1910	Welsh-born English logician and philosopher Bertrand Russell and English mathematician and philosopher Alfred North Whitehead publish the first volume of *The Principles of Mathematics*. They argue that mathematics and logic are identical.
1913	Indian number theorist Srinivasa A. Ramanujan begins a five-year collaboration with English mathematician Godfrey Harold Hardy during which Ramanujan works on and solves many mathematical problems.
1918	Russian trigonometer and educator Nina Bari is the first woman to attend Moscow State University.
1919	Welsh-born English logician and philosopher Bertrand Russell publishes *Introduction to Mathematical Philosophy* which had been largely written while he was in prison for antiwar activities.
1921	German algebraist Emmy Noether publishes her studies on abstract rings and ideal theory which become important in the development of modern algebra.
1921	French number theorist Émile Borel publishes the first in a series of papers on game theory and becomes the first to define games in terms of strategy and to consider the different types of strategies there could be.
1922	German American algebraic geometer Richard Courant founds Göttingen's Mathematics Institute in Germany, making it an international center of theoretical and applied mathematics.

1914
World War I begins.

1917
V. I. Lenin leads communist takeover of Russia.

1923
Edwin Hubble identifies galaxies beyond the Milky Way.

1910   1913   1916   1919   1922

Milestones in the History of Mathematics

1925   African American pure mathematician Elbert F. Cox is the first black to earn a Ph.D. in mathematics when he receives his degree from Cornell University.

1931   Austrian American mathematician Kurt Gödel publishes a paper whose incompleteness theorem startles the mathematical community. It states that within any rigidly logical mathematical system there are propositions that cannot be proved or disproved on the basis of the axioms within that system.

1933   Hungarian number theorist Paul Erdös discovers a proof for Chebyshev's theorem, which says that for each integer greater than one, there is always at least one prime number between it and its double.

1936   Chinese American geometrist Shiing-Shen Chern begins working with French number theorist Elie-Joseph Cartan on differential geometry.

1936   German American algebraic geometer Richard Courant creates a center of mathematics and science at New York University.

c. 1936   Dutch graphic artist M. C. Escher begins to use what could be described as his own version of a mathematical approach to produce his drawings, resulting in work that is part fantasy and part logic.

1937   American mathematician Claude E. Shannon arrives at a connection between a computer's relay circuit and Boolean algebra.

1928
Alexander Fleming
discovers penicillin.

1933
Nazis take control
of Germany.

1937
J. R. R. Tolkien
publishes *The Hobbit.*

1925   1928   1931   1934   1937

1937	English algebraist and logician Alan Turing envisions an imaginary machine that would solve all computable problems and help prove the existence of undecidable mathematical statements.
1943	American mathematician and computer pioneer Howard Aiken completes the Harvard Mark I, the first large-scale automatic digital calculator. Among the scientists at Harvard working on the project is American computer scientist Grace Hopper.
1943	English algebraist and logician Alan Turing helps the World War II allies crack German codes.
1944	Harvard scientists, including American computer scientist Grace Hopper, build the Mark I, the world's first digital computer.
1944	Hungarian American number theorist John von Neumann and Austrian American economist Oskar Morgenstern develop a mathematical theory of games that comes to be known as game theory.
1944	African American mathematical physicist J. Ernest Wilkins Jr. begins a two-year stint at the University of Chicago, working on the Manhattan Project (the code name given to the American effort to build the first atomic bomb).
1945	Hungarian American number theorist John von Neumann presents the first description of the concept of a stored computer program.

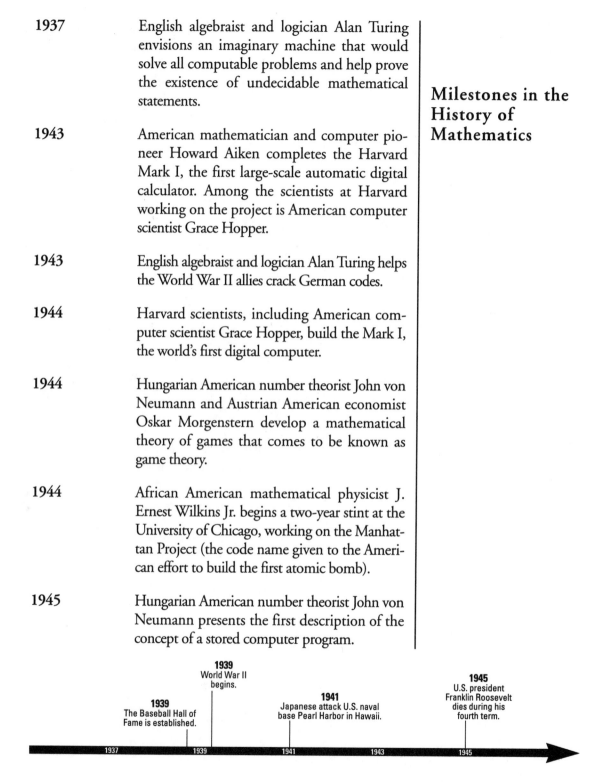

1939
World War II begins.

1939
The Baseball Hall of Fame is established.

1941
Japanese attack U.S. naval base Pearl Harbor in Hawaii.

1945
U.S. president Franklin Roosevelt dies during his fourth term.

1937   1939   1941   1943   1945

1946 The ENIAC (Electronic Numerical Integrator and Computer), the first general-purpose electronic digital computer, is dedicated. Built by two Americans, computer engineer J. Presper Eckert and physicist and engineer John W. Mauchly, the ENIAC would influence all other electronic computers that would follow. Five years later, they completed the UNIVAC (Universal Automatic Computer), a computer for the everyday business world.

1946 The late Danish mathematician Agner K. Erlang is honored when the "Erlang unit" is adopted internationally to mean the total traffic volume of one hour.

1947 African American statistician David Blackwell describes "sufficiency," the process of simplifying a statistical problem by summarizing data.

1948 American logician Norbert Wiener produces a landmark paper that marks the beginning of cybernetics.

1949 University of Michigan students Evelyn Boyd Granville and Marjorie Lee Browne become the first African American women to receive Ph.D.'s in mathematics.

1949 American mathematician Claude E. Shannon formulates basic information theory, upon which much of today's computer and communications technology is based.

1951 Fifteen nations found the International Mathematical Union to promote cooperation among

1948
Jews in Palestine form
the State of Israel.

1949
Mao Zedong
becomes first
leader of People's
Republic of China.

1950
The comic strip *Peanuts*
makes its debut.

1946   1947   1948   1949   1950

the world's mathematicians and to more widely disseminate the results of mathematical research.

1953   American mathematician Claude E. Shannon publishes his pioneering work on artificial intelligence.

1953   African American mathematical physicist J. Ernest Wilkins Jr. and Herbert Goldstein publishes their work on the penetration of gamma rays, which is used in the design of nuclear reactors and radiation shielding. Their study on gamma ray penetration eventually becomes widely used for research in space and for other nuclear science projects.

1954   American computer scientist John Backus invents FORTRAN, the first widely used programming language and the forerunner of nearly all contemporary computer languages.

1956   African American mathematician Evelyn Boyd Granville begins working at IBM as a computer programmer.

1960   The metric system is adopted by nearly every country in the world.

1964   American statistician and computer scientist Thomas E. Kurtz and Hungarian mathematician John George Kemeny develop a general-purpose computer language called BASIC, which soon becomes the most widely used language in the world.

1964   African American algebraist and educator Etta Zuber Falconer begins her long teaching career at Spelman College in Atlanta, Georgia.

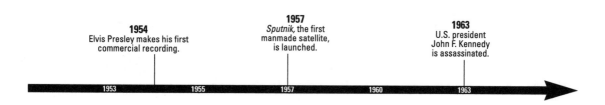

1954
Elvis Presley makes his first commercial recording.

1957
Sputnik, the first manmade satellite, is launched.

1963
U.S. president John F. Kennedy is assassinated.

1953   1955   1957   1960   1963

Milestones in the History of Mathematics

1966

American algebraist Ruth Aaronson Bari receives her Ph.D. from Johns Hopkins University. The graph theory expert goes on to teach for over thirty years at George Washington University, where her work with Ph.D. candidates earns her the nickname "doctoral mother."

1971

American mathematical statistician Mary Gray helps found and is the first president of the Association for Women in Mathematics.

1973

English theoretical physicist and writer Stephen Hawking proves mathematically that what he called mini-black holes actually give off particles and radiation, after which they gradually evaporate and explode. This theory—that black holes are not in fact "black"—has since been accepted by most physicists, and "Hawking radiation" is the term used to describe these black hole emissions.

1978

American mathematician and writer John Allen Paulos combines mathematics and humor to produce his first popular book, entitled *Mathematics and Humor.* The book contains mathematical jokes, riddles, and cartoons.

1980

The work of English applied mathematician Ada Lovelace receives public recognition as the U.S. Department of Defense officially names a software language "Ada" in her honor.

1980

American logician and educator Louise Hay is named the head of the mathematics department at the University of Illinois at Chicago,

1967 The first Super Bowl is played.

1969 Man walks on the moon.

1978 John Paul II becomes pope.

1965　1969　1973　1977　1980

the only woman in the country at the time to head a university department.

1982   Polish-born Lithuanian mathematician Benoit B. Mandelbrot founds fractal geometry, a new branch of mathematics based on the study of the irregularities in nature.

1983   American logician and number theorist Julia Bowman Robinson becomes the first woman to be elected as president of the American Mathematical Society.

1986   Chinese-born American analyst Sun-Yung Alice Chang achieves national recognition when she deliver an address at the International Congress of Mathematicians.

1988   English theoretical physicist and writer Stephen Hawking publishes the best-seller *A Brief History of Time: From the Big Bang to Black Holes.* Intended for a general audience, the book explains in simple language what centuries of people have thought about the nature of the universe and describes the evolution of his own thinking about the cosmos.

1993   English-born mathematician Andrew J. Wiles announces his proof of "Fermat's last theorem." His 200-page paper is the result of a seven-year study on a problem left unsolved by French number theorist Pierre de Fermat 325 years earlier. Over the years, many mathematicians had declared it unsolvable.

Milestones in the History of Mathematics

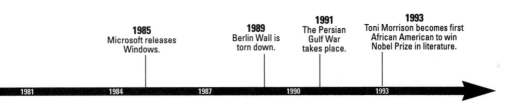

1985
Microsoft releases Windows.

1989
Berlin Wall is torn down.

1991
The Persian Gulf War takes place.

1993
Toni Morrison becomes first African American to win Nobel Prize in literature.

1981   1984   1987   1990   1993

1994	English-born mathematician Andrew J. Wiles publishes a corrected, improved version of his proof of "Fermat's last theorem."
1994	American algebraist and game theorist John Nash wins the Nobel Prize in Economic Science for his early, pioneering work in the field of game theory.
1999	The euro becomes legal tender throughout Europe, beginning a three-year transition to January 1, 2002, when the euro becomes common currency throughout most of Europe.
2001	*A Beautiful Mind,* the successful film about the life of American algebraist and game theorist John Nash, wins the Academy Award for best picture.
2001	American mathematical statistician Mary Gray receives the Presidential Award for Excellence in Science, Mathematics, and Engineering Mentoring.

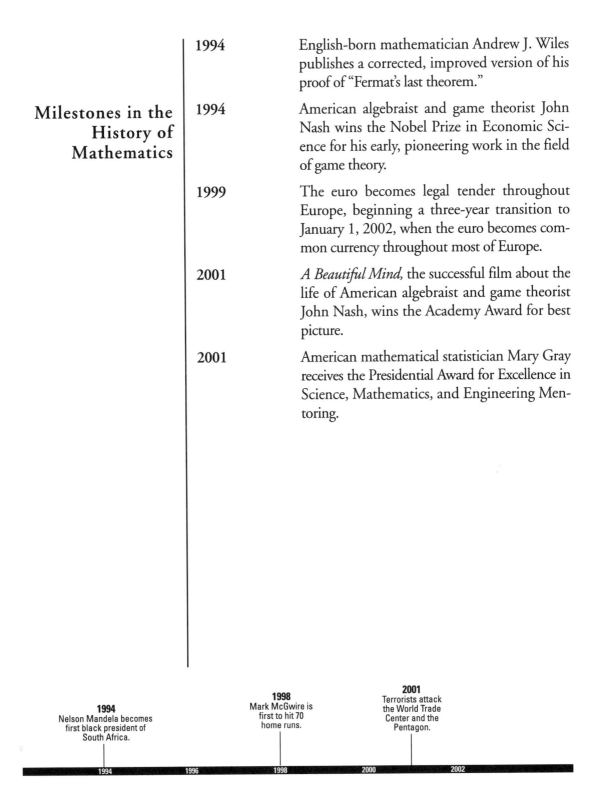

1994
Nelson Mandela becomes
first black president of
South Africa.

1998
Mark McGwire is
first to hit 70
home runs.

2001
Terrorists attack
the World Trade
Center and the
Pentagon.

1994   1996   1998   2000   2002

Born March 8, 1900
Hoboken, New Jersey

Died March 14, 1973
St. Louis, Missouri

American mathematician and computer pioneer

Howard Aiken

Howard Aiken designed and built the Harvard Mark I, the first large-scale automatic digital calculator. This machine and its later, improved versions became the model for more advanced, modern digital **computers** (see entry in volume 1). Aiken was also responsible for introducing computer science as an academic program.

Howard Aiken.
Reproduced by permission of the Corbis Corporation.

From the gas company to Harvard

Howard Hathaway Aiken was born in Hoboken, New Jersey, but was raised in Indianapolis, Indiana, after his parents, Daniel H. Aiken and Margaret Emily Mierisch, moved there. His family's limited resources forced Aiken to go to work when he had completed the eighth grade, and he found a job as a switchboard operator for the Indianapolis Light and Heat Company. During the day, Aiken attended Arsenal Technical High School, then at night he worked twelve-hour shifts. He did this for seven days a week until a school official found out exactly what his work and study schedule was and arranged for a series of tests that allowed Aiken to graduate early. This enabled him to earn more money before applying to the University of Wisconsin at Madison, which he entered in 1919.

Howard Aiken

There, he studied electrical engineering and supported himself by taking a part-time job at the Madison Gas and Electric Company.

After receiving his bachelor of science degree in electrical engineering in 1923, he worked full-time at Madison Gas and Electric where he was promoted to chief engineer. In 1927, Aiken left the company and moved to Chicago, Illinois, where he took a job as an industrial engineer with the much larger Westinghouse Electric Manufacturing Company. After four years, he decided to return to a school environment and in 1931 he accepted a research position in the physics department at the University of Chicago. After briefly going into business for himself and trying out teaching at the University of Miami, Aiken took a career turn that would determine the rest of his life. After deciding in 1935 at the age of thirty-five that he wanted to return to school to work on his Ph.D. in physics, he began his graduate studies at the University of Chicago and soon transferred to Harvard University. In 1937, Aiken received his master's degree in physics from Harvard and was made an instructor there. He then wrote his dissertation while teaching, and received his doctorate in 1939, also in physics.

Proposes design for first modern computer

As with any graduate student in physics, Aiken was required to perform many hours of long, tedious calculations, and it was during these years of hard work that he first began to think seriously about how to speed up this monotonous calculating process. He knew that in the past, others had attempted to build a machine that would do this hard labor quickly, and he was especially influenced by the pioneering, but unsuccessful, work of English mathematician and inventor **Charles Babbage** (1791–1871; see entry in volume 1), who developed many of the basic ideas of modern computers. Since Babbage's time, however, the only things available were electro-mechanical calculating machines that were built for accounting purposes. These accounting machines had none of the calculating power needed by a scientist or engineer, and Aiken soon realized that he would have to invent such a machine himself. Using his experience in mathematics, engineering, and physics, Aiken devoted himself to thinking through this problem, and in 1937, while still a graduate student at Harvard, he produced a twenty-two-page memorandum that contained an original design for his computer.

Herman Hollerith's Punched Cards

Herman Hollerith.
Reproduced by permission of the Corbis Corporation.

When Harvard University approached IBM with Howard Aiken's ideas for a computer, it showed that company a plan that used off-the-shelf technology that had been pioneered by IBM's own founder, Herman Hollerith (1860–1892), some fifty years before.

Hollerith was an American engineer and inventor whose electromechanical punched card sorter and reader conducted the first mechanized census in the United States in 1890. His reader machine had metal pins that passed through holes punched in cards the size of dollar bills, momentarily closing electric circuits. The resulting pulses advanced counters that were assigned to details such as income and family size. His sorter machine was also programmed to pigeonhole cards according to patterns of holes, which proved to be an important aid in analyzing census statistics. Hollerith's automatic sensing made it possible to classify and count data in one-third the time. This marked the real beginning of modern data processing, as Hollerith's machine was eventually adopted all over the world.

In 1911, Hollerith joined two companies to form the Computing-Tabulating-Recording Company which later evolved into IBM. It would be Aiken's invention, however, that would turn IBM away from calculating machines and into a computer giant.

Aiken's plans called for using mostly existing hardware, especially an adaptation of the punched card machines developed by American engineer and inventor Herman Hollerith (1860–1929; see box), which a computing company called International Business Machines (IBM) was using. Although Aiken's machine was to be powered by electricity, its main components would be primarily electromechanical, meaning that it used mechanical parts that are electrically controlled by magnetically operated switches. This meant that it would have to use thousands of electromechanical relays instead of the more modern electric vacuum tubes. (Electromechanical relays use a current flowing in one circuit to open or close a switch in another circuit.) Aiken tried to interest private industry in his plans with no success, but eventually his brief memorandum was promising enough for his colleagues and superiors at Harvard to contact IBM to see if the company would be interested in funding this project.

Builds the Mark I

Aiken's ideas impressed IBM enough that they were willing to make a deal. In 1939, IBM president Thomas Watson Sr. (1874–1956) and Aiken signed a contract stating that IBM would build the computer (with some additional funding from the U.S. Navy), and Aiken would serve as head of the construction team. At this time, IBM only built office machines, but its management wanted to encourage research in new and promising areas, and it was also eager to establish a connection with Harvard University. Eventually, IBM agreed that after the machine was built, it would donate it to Harvard, and IBM would get the credit for building it. In this way, Aiken finally received the support he needed. Working with three IBM engineers, Aiken began work in May 1939 at an IBM facility in Endicott, New York.

It took until January 1943 for Aiken to complete the huge machine. Some of this time was lost because World War II (1939–45) had broken out and everyone's priorities had suddenly changed. Aiken was by then an officer in the U.S. Navy, but he was allowed to keep working on his project. When the machine was finished, it became known officially as the IBM Automatic Sequence Controlled Calculator, or the ASCC, but nearly everyone called it simply the Mark I. What Aiken produced was a huge machine composed of seventy-eight calculators and adding machines. He designed it so that each calculator would work on a part of the same problem, all under the guidance of a single control unit. Aiken chose to make his machine electromechanical because he did not like using vacuum tubes, which he thought were unreliable and inefficient. Because of this, his computer required thousands of relays and other components, all assembled in a huge 51-foot-long, 8-foot-high machine. Since Watson wanted the machine to look futuristic, he had the entire machine covered with a streamlined case of glass and steel.

Because it was primarily electromechanical, everything about the Mark I was large. It used some five hundred miles of wire, contained more than three million connections, and weighed thirty-five tons. While Mark I was faster than any previous computer, it

was relatively slow because it was not all-electronic. While it could perform calculations up to twenty-three digits long, it still took six seconds to perform multiplication, and twelve seconds to divide two numbers. Performing all of these calculations meant that its rotating shafts, clutches, switches, and relays were all spinning, opening, and closing constantly. Since relays are relatively large and make a loud noise when they open or close, the Mark I made quite a racket, and was described as sounding like a "long room full of [knitters] with steel needles."

The Mark I did not resemble a modern computer, either in appearance or in the way it worked. It had no keyboard, but instead received its instructions from punched paper tape. Input was entered from punched cards or tape or from hand-set switches, and output was on punched cards or printed on paper. The U.S. Navy put the Mark I to work during World War II, and it performed all the complex mathematical calculations involved in

Howard Aiken stands in front of his Mark I computer, which was completed in January 1943. *Reproduced by permission of the Corbis Corporation.*

aiming long-range guns that were onboard a ship. It was reliable and was always up to the job it had to do.

Builds Mark II, III, and IV

Following the success of the Mark I, Aiken went on to build three more advanced versions. The Mark II, completed in 1947, was a fully electronic machine. The Mark III was completed in 1950, and Mark IV was ready in 1952. All were used by the U.S. military, and each was faster than the other. Although all three versions were faster than Mark I machines and used electronic components such as vacuum tubes and solid-state transistors, they often broke down and were never as reliable as the Mark I.

The most important factor against the Mark computers, however, was that they did not have the capability for what came to be known as a "stored program." Having a stored program meant that special instructions for each job did not have to be loaded each time the computer was run. It was on this issue that Aiken would put himself out of step with any future progress. For although Aiken was in many ways a visionary and a true computer pioneer, his vision did not include the concept of a stored program, about which he had serious doubts. All future computers would eventually come to have some version of a stored program. Aiken was also extremely conservative about society's future need for computers, predicting in 1947 that only six electronic computers would be needed to satisfy the computing requirements of the United States. However, this was not totally outside the thinking of many in society at the time, as nearly everyone believed that in the future, computers would be used only for large scientific calculations or data processing for the government, large industry, or large educational and research establishments.

Although Aiken was quickly surpassed by many other people whose names are sometimes better known, he performed an invaluable service to early computing by demonstrating that a large calculating computer could not only be built, but could also provide the scientific world with high-powered, relatively speedy mathematical solutions to any number of complex problems. Aiken stayed at Harvard until 1961, having introduced computer science as an academic program there. Some say that this may have been his greatest contribution. After he retired from Harvard and moved to Fort Lauderdale, Florida, in 1961, Aiken went on to help the University of Miami set up a computer science program and a computer center there as well. During his lifetime, Aiken received several major awards, and is thought of by many as the father of modern computers. He died in Missouri at the age of seventy-three.

Howard Aiken

For More Information

Abbott, David, ed. *Mathematicians.* New York: Peter Bedrick, 1985.

Ferguson, Cassie. "Howard Aiken: Makin' a Computer Wonder." *Harvard University Gazette* (April 9, 1998). http://www.news.harvard.edu/gazette/1998/04.09/HowardAikenMaki.html (accessed July 3, 2002).

Freed, Les. *The History of Computers.* Emeryville, CA: Ziff-Davis Press, 1995.

"Howard Aiken's Harvard Mark I (the IBM ASCC)." *A History of Computers.* Maxfield & Montrose Interactive. http://www.maxmon.com/1939ad.htm (accessed July 3, 2002).

Mathematicians and Computer Wizards. Detroit: Macmillan Reference USA, 2001.

Ritchie, David. *The Computer Pioneers: The Making of the Modern Computer.* New York: Simon and Schuster, 1986.

O'Connor, J. J., and E. F. Robertson. "Howard Hathaway Aiken." *The MacTutor History of Mathematics Archive.* School of Mathematics and Statistics, University of St. Andrews, Scotland. http://www.groups.dcs.st-andrews.ac.uk/~history/Mathematicians/Aiken.html (accessed July 3, 2002).

Born c. 250 B.C.E.
Perga, Asia Minor (now part of Turkey)

Died c. 180 B.C.E.
Alexandria, Egypt

Greek geometer

Apollonius of Perga

Apollonius of Perga was the last of the great Greek mathematicians. Called the "Great Geometer" by his contemporaries, he originated many of the geometric shapes and terms that would become key to the later development of **geometry** (see entry in volume 1), astronomy, mechanics, and navigation. He is also considered to be one of the founding fathers of mathematical astronomy in ancient Greece.

Few details of personal life

So little is known about the life and times of Apollonius of Perga that scholars are not even sure of the years he was born and died. Scholars have compiled some facts from what others have written about him and from what he wrote himself. Still, most of what is said about his life should always be prefaced by the word "probably." He is said to have been born in Perga, a small Greek city in southern Asia Minor that is now a part of Turkey. Around the second half of the third century B.C.E., when he is thought to have been born, Perga was both a major center of culture and learning and the center of worship of Queen Artemis (pronounced ARE-teh-miss), the goddess of nature. Apollonius is thought to have been the son of a geometer, and as a young man, studied in Alexandria with the pupils of Greek geometer and logician **Euclid of Alexandria** (c. 325–c. 270 B.C.E.; see entry in volume 1). Researchers also know Apollonius studied at the newly built library and university in Pergamum (pronounced PER-guh-mum), then a major cultural center, also in what is now Turkey near the Aegean (pronounced ee-JEE-un) Sea.

In the prefaces of some of his books, Apollonius left details that scholars have used to piece together further details of his life. Researchers know that he visited the city of Ephesus (pronounced EH-fuh-suhs), that he had a son named Apollonius, and that he addresses one book to a person named Eudemus (pronounced YOO-duh-muss) and another to Atallus (pronounced EH-tuh-lus). Finally in one preface, Apollonius mentions Philonides (pronounced fih-low-NYE-dees) the geometer, "whom I introduced to you in Ephesus." Since scholars know when Philonides lived, they are better able to place Apollonius in time. Researchers also know that he taught in Alexandria. Other Greek writers who lived long after he did wrote about him. For example, Greek mathematician Pappus (pronounced PAH-pus) of Alexandria (c. 290–c. 350 C.E.) wrote that Apollonius was vain, a braggart, and jealous of others. However, readers should take into consideration that Pappus was writing about a person who lived some five hundred years before him.

Treatise on the conics

Apollonius is best known and remembered today because of his study of certain geometric figures known as conic (pronounced KAHN-ik) sections. They were called conic sections because they were, in fact, sections or pieces that resulted when a cone was cut at different angles. In his major work, titled *The Conics*, Apollonius offered new ideas on how to subdivide a cone to produce **circles** (see entry in volume 1). *The Conics* is divided into eight books containing about four hundred logical statements. The first four books consist of an introduction and an account of the current state of mathematics and also give a survey of Greek geometry. The last four books contain his highly original work on conic sections. However, only books one through four survive in the original Greek language, while books five through seven exist in Arabic. No version of his book eight exists, although several mathematicians have tried to reproduce what they thought it would contain.

Apollonius of Perga

In books five through seven, Apollonius shows how, by cutting through cones at particular angles, he could produce entirely new geometric figures. He then described these new shapes, and even gave them names such as ellipse (pronounced ee-LIPS; a shape resembling a flattened circle), parabola (pronounced puh-REH-bow-luh; a bowl shape), and hyperbola (pronounced hye-PUR-bow-luh; a type of curve). He then investigated the properties of each of these new shapes and showed how they were all interrelated.

Apollonius's eight books on conic sections were so original and complete that they soon became the accepted and unquestioned authority in their field. Much later, during the scientific revolution of the seventeenth century, his work became essential to the physics and astronomy of English mathematician and physicist **Isaac Newton** (1642–1727; see entry in volume 2). As noted in Julia Diggins's *String, Straightedge, and Shadow: The Story of Geometry*, Newton may have been referring to Apollonius when he stated, "If I have seen further than other men, it is because I stood on the shoulders of giants."

In more modern times, the work of Apollonius has proven to be a milestone in the understanding of mechanics, navigation, astronomy, and even space science. This is because conic sections trace the paths travelled by projectile—like bullets, missiles, or planets—and his work is especially relevant to today's ballistics, the study of the motion of projectiles in flight and space science. Much of the astronomy worked out by second-century Greek astronomer **Claudius Ptolemy** (c. 100–c. 170; see entry in volume 3) was based on the ideas of Apollonius concerning epicycles and ellipses. (An epicycle is a circle whose center lies on the circumference or the rim of a larger circle.) Ptolemy placed Earth at the center of the universe and used the notion of epicycles to explain the orbits of the planets. Although his ideas were wrong, they dominated Western astronomy for some fifteen hundred years.

Other than *The Conics,* only one other small essay that Apollonius wrote survives, though it does not carry the importance of *The Conics.* Although Euclid's geometry is perhaps the best known and most popular work of Greek mathematics, scholars believe the original work on conics done by Apollonius of Perga to be the greatest achievement of Greek geometry.

For More Information

Abbott, David, ed. *Mathematicians*. New York: Peter Bedrick Books, 1985.

Diggins, Julia E. *String, Straightedge, and Shadow: The Story of Geometry*. New York: Viking Press, 1965.

Franceschetti, Donald R., ed. *Biographical Encyclopedia of Mathematicians*. New York: Marshall Cavendish, 1999.

Mathematicians and Computer Wizards. Detroit: Macmillan Reference USA, 2001.

O'Connor, J. J., and E. F. Robertson. "Apollonius of Perga." *The MacTutor History of Mathematics Archive*. School of Mathematics and Statistics, University of St. Andrews, Scotland. http://www.groups.dcs.st-andrews.ac.uk/~history/ Mathematicians/Apollonius.html (accessed July 3, 2002).

Toomer, G. J. "Apollonius of Perga." *Biographical Dictionary of Mathematicians*. New York: Charles Scribner's Sons, 1991, pp. 66–80.

Born December 3, 1924
Philadelphia, Pennsylvania

American computer scientist

John Backus

John Backus invented FORTRAN, the first widely used programming language and the forerunner of nearly all contemporary **computer** (see entry in volume 1) languages. As one of the early visionaries in the world of computer science, Backus made possible the practical use of computers as high-speed tools of industry and science.

A confused beginning

John Backus was born in Philadelphia, Pennsylvania, in 1924, but grew up in Wilmington, Delaware. Since his family was well-off, he was sent to a private high school, the Hill School, in Pottstown, Pennsylvania. Backus did not do well there, and often failed his classes. As noted in *The MacTutor History of Mathematics Archive,* he later recalled these difficult teen years saying, "I flunked out every year. I never studied. I hated studying. I was just goofing around." He never minded having to attend summer school, however, since it was located in New Hampshire and he got to go sailing most days. Despite his poor performance in school, he was able to graduate with his class in 1942 and entered the University of Virginia later that year. Backus had planned to major in chemical engineering in college, but his bad high school habits continued and he missed so many classes and did so poorly that he was asked to leave before he finished his freshman year.

Since he was out of school in 1943 and the United States was fully engaged in World War II (1939–45), Backus decided to join the U.S. Army. At this point, his life began to take some interesting turns. While stationed at Fort Stewart, Georgia, and assigned to an antiaircraft crew, he found himself suddenly reassigned because of an aptitude test he had taken. This test indicated that he had a natural ability for engineering, and the Army enrolled him in a pre-engineering program at the University of Pittsburgh. While there, he took a medical aptitude test and did so well on it that he was accepted into the premedical program at Haverford College in Haverford, Pennsylvania, as part of the Army's specialized training. His reassignment back to school by the Army may have saved his life, since his old battalion (pronounced buh-TAL-yun; division) eventually was shipped out to fight in France in what became known as the Battle of the Bulge. This bloody and bitter battle in December 1944 was Germany's final offensive action in France. Back home things got even stranger for Backus as he was assigned by the Army six months of premedical training at a hospital in Atlantic City, New Jersey. While there it was discovered that Backus had a non-cancerous bone tumor in his skull, and after successful surgery, he had a metal plate fitted in his head.

Discovers his love of mathematics and joins IBM

In May 1946, the war was over and Backus was out of the Army. By this time, he had already given up on a medical career. As noted in *The MacTutor History of Mathematics Archive,* Backus remembered that in medical school, "They don't like thinking.... They memorize—that's all they want you to do." Soon, he found himself out of school and renting an apartment in New York City but not knowing what he wanted to do with his life. However, one more strange turn would eventually lead him to a field about which he knew nothing but in which he would become a true pioneer. In New York, he enrolled in a school for radio technicians almost on a whim, and while there discovered that he had a strong interest and ability in mathematics. He later recalled how this happened saying, "I had a very nice teacher—the first good teacher I ever had—and he asked me to cooperate with him and compute

John Backus

the characteristics of some circuits for a magazine." This experience was so enjoyable and satisfying that he decided to enroll in Columbia University in New York to study mathematics.

John Backus

This time, Backus did not neglect his studies in school, and in 1949 he earned a bachelor's degree in mathematics from Columbia. A year later, he received a master's degree, also from Columbia and also in mathematics. Still not knowing how he would use his education, Backus was once again led to a life-changing situation by what seemed like chance. During the spring of 1949, as he was nearing graduation, he had taken a tour of the IBM Computer Center in New York City where a tour guide encouraged him to apply for a job. That same day he applied, did very well on a test, and eventually was offered a job and joined IBM in 1950.

During his first year at IBM, Backus was assigned to work on the Selective Sequence Electronic Calculator (SSEC). Although he knew very little about computers, he learned quickly and soon made a real contribution. The SSEC was very much like every huge first generation computer. It was slow, full of moving parts, had no memory, and broke down too often. As a programmer for the SSEC, Backus studied the machine and soon invented a program called Speedcoding. This successful program enabled the computer to store complex data quickly and efficiently. Speedcoding was also applied with good results to the IBM 701 computer. Backus had made his first real contribution and was quickly building a reputation as a trailblazer.

Develops FORTRAN

By 1952, IBM was keenly aware that its newest computer, the IBM 704, would also need much more advanced programming language than was available, and Backus was assigned to this project. Since Backus had previously written a memorandum outlining his plan for a new, advanced language, he was made the leader of a team of programmers. At this point in time, the operating costs of programming and then debugging (finding the programming errors) a computer were at least as high as the cost of the hardware itself. This was because no practical, high-level programming language was available. With this in mind, Backus and his team at IBM wanted to develop a simple language that scientists and engineers could use to write computer programs. Above all, they wanted to create a language that was easy for nonspecialists to learn and use. The programming language they developed was called FORTRAN, which was an acronym for FORmula TRANslation.

The wonderful thing about what Backus and his team produced was that he not only created a new and simple programming language, but he also provided a mechanism that would serve as the "translator" between the human user and the computer brain. Called a compiler or translator, this device would translate human programming language for a computer whose only language is zeroes and ones. This translating device proved to be a major breakthrough, for it did away with all the time-wasting, tiresome hand coding that also introduced so many errors. The compiler or translator proved to be a truly revolutionary breakthrough in computer speed and efficiency. By 1956, twenty-five thousand lines of FORTRAN machine code on magnetic tape had become standard capacity for every IBM 704, an enormous leap in computer speed. Magnetic tape was one of the earliest forms for storing large quantities of encoded computer data at low cost. It usually came in large, long reels of ½-inch tape, sometimes containing as much as thirty-six hundred feet on a spool.

Amazingly, however, although Backus wrote FORTRAN for one particular IBM computer, it was adopted and adapted to many other systems and soon became the industry standard. Neither Backus nor his colleagues thought that FORTRAN would be used on machines other than the IBM 704, nor did they anticipate that it would remain in use even into the beginning of a new century and be recognized as the most important innovation in the history of programming languages. However, the beauty of his design is that it allows people with no special knowledge of computers, nor with even any idea of how computers work internally, to be able to deal with and use them.

FORTRAN naturally went through several changes that made it better and even more reliable. It helped make such milestone events as the American manned moon landing of 1969 possible, and contributed to many other computer-based breakthroughs in all sorts of fields as well. Throughout the last quarter of the twentieth century and into the start of the twenty-first century, FORTRAN remained the standard language for scientific applications.

John Backus

John Backus

Joins international programming team

In the late 1950s, an international team was created to develop a multipurpose language that would be completely independent of specific computers and that would work with any and all computers. Backus was part of this international programming and design team, and the team soon produced an international algebraic language called ALGOL. Although ALGOL was not a huge success like FORTRAN and did not become widely popular, it did serve as the foundation and strongly influenced the creation of three other major programming languages: Pascal, C, and Ada.

In the 1970s and 1980s, Backus continued to work on ways to improve computer languages and sought an ever more efficient programming language. He received the 1975 National Medal of Science and the 1977 Turing Award from the Association for Computing. In 1991, Backus, who is married to Barbara Una Stannard and has two daughters, retired from IBM; two years later, he received the 1993 Charles Stark Draper Prize, the world's most prestigious engineering award. Retired from the world of computer science and living in California, Backus remains a man for whom mathematics made all the difference in the world. It gave him enthusiasm as well as a purpose and a goal. It allowed him to find and express the greatness within himself that had been undiscovered, which then enabled him to become a real computer pioneer.

For More Information

Franceschetti, Donald R. *Biographical Encyclopedia of Mathematicians.* New York: Marshall Cavendish, 1999.

Mathematicians and Computer Wizards. Detroit: Macmillan Reference USA, 2001.

O'Connor, J. J., and E. F. Robertson. "John Backus." *The MacTutor History of Mathematics Archive.* School of Mathematics and Statistics, University of St. Andrews, Scotland. http://www.groups.dcs.st-andrews.ac.uk/~history/Mathematicians/Backus.html (accessed July 8, 2002).

Shasha, Dennis Elliott, and Cathy Lazere. *Out of Their Minds: The Lives and Discoveries of 15 Great Computer Scientists.* New York: Copernicus, 1995.

Weiss, Sonia. "John Backus, Inventor of FORTRAN." *Jones Telecommunications & Multimedia Encyclopedia.* http://www.digitalcentury.com/encyclo/update/backus.html (accessed July 8, 2002).

John Backus

Born November 19, 1901
Moscow, Russia

Died July 15, 1961
Moscow, Soviet Union

Russian trigonometer and educator

Nina Bari

Nina Bari was a major contributor to the study of functions and sets, and was a turn-of-the-century pioneer in these more abstract areas of mathematics. As a woman mathematician in the Soviet Union, she was the first of her gender in that country to succeed in a male-dominated field. After her tragic death, she was remembered not only as an outstanding research mathematician but as someone who really enjoyed teaching mathematics.

The Bolshevik revolution drastically changes Russia

Nina Karlovna (pronounced car-LOHV-nuh) Bari was born in Moscow, Russia, the daughter of Karl Adolfovich Bari, a doctor, and Olga Eduardovna. As a member of Russia's educated elite, she was sent to a private high school for girls, where she demonstrated great mathematical ability. In those days, however, Russia educated boys and girls separately, and she could not take advantage of the best teachers who were usually sent to boys' schools. This situation changed dramatically after 1917 when the Russian Revolution of 1917 broke out. After the ruling family of Russia finally gave up its power in the spring of 1917, several different revolutionary groups struggled to take charge of the country. By November of that year, the revolutionary group known as the Bolsheviks (pronounced BOWL-shuh-vix) finally seized power and ended the brief civil war. During 1918, this group overpowered any opposition and created a new one-party Communist state. This one-party dictatorship eventually eliminated private property and instituted many sweeping changes throughout the land, including changing the name of the country from Russia to the Union of Soviet Socialist Republics, also known as the Soviet Union.

Pioneers in a man's field

Although the Communist takeover of Russia resulted in many misdeeds, injustices, and even crimes, it had a short-term positive effect on education. Many educators were now radical or at least progressive thinkers who felt that everyone should be able to receive an education. Thus, when the Faculty of Physics and Mathematics of Moscow State University reopened in 1918, it began accepting applications from women. Records show that Bari was the first woman to attend Moscow State, which, throughout its long history, had been an institution of higher learning for men only. During these early revolutionary years, the educational institutions of the Soviet Union were in the same turmoil as the society around them, and graduation examinations were scheduled only occasionally. Bari asserted herself and took advantage of this disorder to sit for her examinations early. Thus, she graduated earlier than normal—one year—and probably became the first woman to graduate from Moscow State.

Finds a mentor and a movement

While at Moscow State, Bari became a member of several mathematical groups, but she was especially drawn to a select group named the "Luzitania" (pronounced loo-zih-TAY-nee-uh). Students in this group followed the mathematical ideas of Nikolai Nikolaevich Luzin (1883–1950), a noted practitioner of the "new" mathematics. Luzin was an inspirational teacher who had only one goal in mathematics—to study nothing but function theory. The idea of "function," as it is used in mathematics, is a simple concept that turned out to be one of the most useful in mathematics, although most of its applications are found in higher math and are well beyond beginning **algebra** (see entry in volume 1). A

Nina Bari

function is the connection between two or more variables. In other words, function always means that there is a relationship between things. For example, if two variable or changeable quantities x and y are related so that, when x is given a value, there is a corresponding value of y, then y is called a function of x. In the solution to the well-known time and distance problem (in which distance equals rate times time or d=rt), if the rate something is travelling is known, then time is a function of rate.

Since Luzin was worshipped by Bari and his other followers, his scholarly goals became theirs as well. The idea of function became especially important to trigonometry (mathematics that deals with the sides and angles of triangles), and Bari, like her mentor Luzin, would devote her life's work to this one aspect of mathematics. She made function theory the major subject of her research at Moscow State and developed her thesis around this topic.

Travels abroad and establishes herself

After graduating early from Moscow State, Bari began her teaching career and lectured at three different institutes before obtaining a fellowship at the newly opened Research Institute of Mathematics and Mechanics at Moscow State. There, she taught and began work on her doctorate under Luzin. She also continued her work on functions and in 1922 presented some of her results to the Moscow Mathematical Society, becoming the first woman to address this society. A year later, she published her first paper, and in 1926, she was awarded her doctorate in mathematics, winning a major prize in the process. In the spring of 1927, she was given the opportunity to study outside of the Soviet Union, and went to Paris, France, to study at the Sorbonne and the Collège de France, where she studied under some of the leading French mathematicians. She then attended the Polish Mathematical Congress in Lvov (pronounced luh-VUHV), Poland, and went to Bologna (pronounced bowl-LONE-yuh), Italy, in 1928, where she lectured at the International Congress of Mathematicians. A Rockefeller grant allowed her to return to Paris and study for another year. In 1932, Bari was made a full professor at Moscow State, and by 1935, her lectures and many publications had established her internationally as a leading mathematician in the theory of functions. That same year, she received another degree, doctor of the physical-mathematical sciences, which is a higher research degree than a traditional Ph.D. During the 1940s and 1950s, Bari continued to devote all her time to teaching at Moscow State, which is what she really loved to do. Her enthusiasm always carried over to her students and she was a very popular professor.

Personal life

Along with her great love and enjoyment of mathematics, Bari had many other interests and pleasures. She loved to read great literature and was known to write very well herself. She also greatly enjoyed music, the ballet, and all of the arts, but was most passionate about hiking in the wilderness. During her early years at Moscow State, she met Soviet mathematician Viktor Vladimirovich Nemytskii, who was one year older than she. Nemytskii was not only a mountain hiker, but a true explorer, and she accompanied him on many of his difficult and dangerous excursions through the country's unexplored mountains. In fact, one of the passes in the Caucasus (pronounced KAW-kuh-suhs) region of Europe (covering parts of Azerbaijan, Armenia, Georgia, and Russia) is named after him. Most believe that he and Bari married at some point, although there is no documentation to prove it.

On July 15, 1961, Bari was killed suddenly after she fell in front of a train at the Moscow Metro. Just before her death, she had published a nine-hundred-page book in her field, which was her fifty-fifth publication. Some say her death was not an accident but rather a suicide, as she was depressed over the death, some ten years before, of her mentor (and some say her lover), Luzin. However, no one knows for sure. At the time of her death, she was considered to be the principal leader of mathematics at Moscow State University. As noted in *The MacTutor History of Mathematics Archive,* one of her students viewed her death as "a great loss for Soviet mathematics and a great misfortune for all who knew her. The image of Bari as a lively, straightforward person with an inexhaustible reserve of cheerfulness will remain forever in the hearts of all who knew her."

For More Information

Franceschetti, Donald R., ed. *Biographical Encyclopedia of Mathematicians.* New York: Marshall Cavendish, 1999, pp. 39–41.

Nina Bari

O'Connor, J. J., and E. F. Robertson. "Nina Karlovna Bari." *The MacTutor History of Mathematics Archive.* School of Mathematics and Statistics, University of St. Andrews, Scotland. http://www.groups.dcs.st-andrews.ac.uk/~history/Mathematicians/Bari.html (accessed July 8, 2002).

Soublis, Giota. "Nina Karlovna Bari." *Biographies of Women Mathematicians.* http://www.agnesscott.edu/lriddle/women/bari.htm (accessed July 8, 2002).

Spetich, Joan, and Douglas E. Cameron. "Nina Karlovna Bari." In *Women of Mathematics: A Bibliographic Sourcebook.* Edited by Louise S. Grinstein and Paul J. Campbell. Westport, CT: Greenwood Press, 1987, pp. 6–12.

Born March 18, 1870
Halifax, Nova Scotia, Canada

Died March 9, 1917
Columbia, Missouri

Canadian algebraist

Agnes Sime Baxter

A gnes Baxter was among the first generation of women in North America to earn a Ph.D. in mathematics. She was the second woman in Canada to earn a doctoral degree in mathematics, and the fourth in all of North America to do so. Despite her abilities, Baxter gave up mathematics to help support her husband's career and raise a family, and she died before she could ever do any real mathematical work again.

A distinguished and honored student

Agnes Sime Baxter was born in Halifax, Nova Scotia, which is on the northeast coast of Canada, opposite the north coast of the state of Maine, and is one of Canada's four maritime or Atlantic provinces. Little is known about Baxter's early life, but scholars do know that her father's name was Robert Baxter, and that Agnes enrolled at Dalhousie University in Halifax at the age of seventeen. Baxter majored in mathematics at Dalhousie and was part of a mostly male class. In 1891, she received her bachelor's degree with "first class distinction," the first female student at Dalhousie given such an honor. That same year, she was awarded the Sir William Young Gold Medal in Mathematics. As a top mathematics stu-

Agnes Sime Baxter

dent, Baxter remained at Dalhousie and earned her master's degree in mathematics and physics in 1892. After applying for and winning a scholarship, she attended Cornell University in Ithaca, New York, and was awarded her Ph.D. in mathematics in 1895.

Cornell a friendly place

During the nineteenth century, Cornell University was one of the few American educational institutions to show any leadership in educating women. Part of this stemmed from the very real interest of its founder, American inventor and businessman Ezra Cornell (1807–1874). Cornell opened Cornell University in 1868, and by the 1890s, the school had become an inviting place for women who were seeking to obtain a higher education. This was especially so for Cornell's mathematics department, which granted three of the first six Ph.D.'s earned by women in the United States. Baxter was the third of these women to earn her Ph.D. in math, as well as the fourth in all of North America (which includes the United States, Canada and Mexico). Unfortunately for Baxter and many of her female colleagues, however, few of these talented women were able to realize their full potential by having a career in mathematics, since most academic positions were given to men.

Abandons mathematics

The postgraduate experience that Baxter encountered was fairly typical of what often happened to many women mathematicians around the turn of the twentieth century. Just one year after receiving her Ph.D., Baxter decided to marry a fellow student, A. Ross Hill. Like Baxter, Hill was also a graduate of Dalhousie University and he received his Ph.D. in philosophy from Cornell the same year that Baxter earned hers in mathematics. The couple soon started a family, and when Hill accepted the position of president of the University of Missouri in 1903, Baxter stopped pursuing her interest in mathematics and decided to fully support her husband's career. As the wife of a university president, this meant hosting school functions and helping to raise funds instead of writing mathematical papers or teaching at the college level. Sadly, Baxter developed a slow, wasting illness that kept her in poor health for many years. After a long medical struggle when it seemed that she might be cured, she suddenly became even sicker and finally died of an infection at the age of forty-seven.

John Charles Fields

John Charles Fields.
Reproduced by permission of
Mathematisches Forschungsinstitut
Oberwolfach.

While Agnes Baxter is probably Canada's best-known woman mathematician, there is little doubt that Canada's best-known male mathematician is algebraist John Charles Fields (1863–1932). A contemporary of Baxter's, Fields received his bachelor's degree from the University of Toronto in 1884 and his Ph.D. from Johns Hopkins University in Baltimore, Maryland, in 1887. After studying in Paris and Berlin, Fields returned to his native Canada and became the first significant research mathematician at the University of Toronto. He would remain on that school's faculty his entire career. Although he was known for his elegant and original mathematical research, Fields is best remembered for establishing the Fields Medal, regarded by all as the Nobel Prize of mathematics.

As the organizer and president of the 1924 International Congress of Mathematicians (ICM), Fields proposed that a special international prize be given to young research mathematicians who had already done significant work. Years before, in 1896, when Swedish inventor Alfred Bernhard Nobel (1833–1896) died, his will provided funds for establishing annual prizes in several fields. Since the inventor of dynamite loved physics and chemistry, these two fields were a natural, as were medicine and peace since they benefited all mankind. Since he also greatly loved literature, one was established for that subject. But Nobel is said to have so strongly disliked Swedish mathematician Gösta Mittag-Leffler (1846–1927), that he would not create a prize for mathematics. Although this may not be true, it is known that Nobel was an extremely practical man who did not care very much for basic research or theoretical ideas. When Fields fell ill in 1932 and the ICM still had not adopted his proposal, he dictated a will that donated his estate to the ICM to be used for the prize. That same year, Fields died and the ICM created the International Medal for Outstanding Discoveries in Mathematics. Everyone immediately called it the Fields Medal, however.

The first Fields Medal was given in 1936, and since 1950 it has been given every four years. The award is restricted to mathematicians not over forty, and up to four medals may be awarded each time. The actual prize is $15,000, and the medal is gold-plated and 11 inches in diameter. It contains the head of **Archimedes of Syracuse** (287–212 B.C.E.; see entry in volume 1) on its front, surrounded by the Latin words "Transire suum pectus mundoque potir," meaning "To rise above human limitations and grasp the world." The prize awards both work completed and potential for future achievement. No woman mathematician has ever won the Fields Medal.

Agnes Sime Baxter

A tribute and honor in death

After Baxter's death, her husband thought that he could best serve Baxter's memory by making a gift of $1,000 to Dalhousie University so that it might create a fund to buy needed books. At the time of her death, the death notices that appeared all mentioned Baxter's considerable ability in the difficult (and usually male-dominated) field of mathematics. After her death, her husband made it clear that his gift to Dalhousie was intended as a form of recognition to her for supporting his career. As noted in *Biographies of Women Mathematicians Web Site*, Hill said he was making the gift, "to perpetuate the memory of one of its loyal graduates, who gave her life to assist in my educational work instead of making an independent record for herself." Obviously, Baxter's husband was well aware of her capabilities in mathematics and he knew she could have pursued her own career.

On March 5, 1988, Dalhousie University dedicated the Agnes Baxter Reading Room in the Department of Mathematics, Statistics and Computing Science. This reading room houses the largest magazine collection in mathematics and statistics in Canada's maritime provinces. The gold medal that Baxter won as an undergraduate at Dalhousie in 1891 is also on display there. Baxter's short life was one of unrealized potential as a scholar, but hopefully, the personal satisfaction she realized with her husband and family made up for the intellectual and professional heights that she never allowed herself to achieve.

For More Information

"Cochell: The Early History of the Cornell Mathematics Department." Cornell University. Department of Mathematics. http://www.math.cornell.edu/General/History/historyP7.html (accessed July 8, 2002).

Lòpez-Ortiz, Alex. "Why Is There No Nobel in Mathematics?" *Frequently Asked Questions in Mathematics*. http:// www.cs.unb.ca/~alopez-o/math-faq/mathtext/node21.html (accessed July 8, 2002).

O'Connor, J. J., and E. F. Robertson. "Agnes Sime Baxter." *The MacTutor History of Mathematics Archive*. School of Mathematics and Statistics, University of St. Andrews, Scotland.

http://www.groups.dcs.st-andrews.ac.uk/~history/
Mathematicians/Baxter.html (accessed July 8, 2002).

O'Connor, J. J., and E. F. Robertson. "John Charles Fields." *The MacTutor History of Mathematics Archive.* School of Mathematics and Statistics, University of St. Andrews, Scotland. http://www.groups.dcs.st-andrews.ac.uk/~history/ Mathematicians/Fields.html (accessed July 8, 2002).

Riddle, Larry. "Agnes Baxter." *Biographies of Women Mathematicians Web Site.* http://www.agnesscott.edu/lriddle/women/ baxter.htm (accessed July 8, 2002).

Agnes Sime Baxter

Born 1114
Biddur (now Bidar), India

Died c. 1185
Ujjain, India

Indian algebraist and astronomer

Bhāskara II

Bhāskara II was the leading mathematician of the twelfth century and is probably the best known mathematician of ancient India. A true master at mathematics, he wrote the first work that fully and systematically used the **decimal** (see entry in volume 1) number system. He also used letters to represent unknown quantities, as in modern **algebra** (see entry in volume 1).

Descendant of a long line of scholars

Bhāskara (pronounced BAAS-kuh-ruh) is often called Bhāskara II to distinguish him from a much earlier Indian mathematician named Bhāskara (himself, often called Bhāskara I), who lived about five hundred years earlier. He is also sometimes called the Bhāskara the Learned or Bhāskarācārya. Born in Biddur (now Bidar) in southern India, he was related to Indian mathematician and astronomer Brahmagupta (598–c. 665; pronounced bruh-muh-GOOP-tuh), and his father was an astrologer named Mahesvara. Bhāskara II must have learned mathematics from his father since in Indian society of the time, the educated males taught their own sons and other family members. It is likely Mahesvara recognized his son's genius at a very early age for

Bhāskara II would go on to become the most original mathematician of his age.

Heads the astronomical observatory

Eventually, Bhāskara II would become the head of the astronomical observatory in the city of Ujjian (pronounced OOH-jyne). Located in central India, this ancient city is considered by Hindus to be one of the country's holiest places. Hindus practice Hinduism, which is the dominant religion of India. Ujjain was also the leading mathematical center of ancient India, so it is fitting that it was there that Bhāskara II composed his most important books on mathematics. In linking mathematics and astronomy, Bhāskara II wrote as many as six works, some of them far more advanced than the work of his time. In his *Līlāvatī* (The Beautiful), which he addressed to either his wife or his daughter, he used decimal notation or the decimal number system in a full and systematic way. Decimals are like **fractions** (see entry in volume 1) in that they are used to represent part of a whole. They also have sometimes been described as another way of writing a fraction whose denominator is some power of ten. Although Bhāskara II did not invent decimal notation, he gave it its first full use, and *Līlāvatī* is considered the first known published work that uses decimal notation. In Europe, the decimal system would not receive widespread use until the sixteenth century.

Other important works

Bhāskara II also wrote a book that focuses on algebra called *Bījaganita* (Seed Counting or Seed Arithmetic). It is significant that in this work he used letters to represent unknown quantities, much as in modern algebra. His masterpiece is a two-part work on mathematical astronomy titled *Siddhāntaś iromani* (Crown of Accuracy). The mathematical part of this work is full of new methods, techniques, and ideas, such as his anticipation of the modern use of signs. Near the end of his life, Bhāskara II published a work on astronomy titled *Karanakutāhalam* (The Calculation of Astronomical Wonders), which popularized the astronomical knowledge of his time.

While little is known of his personal life, scholars do know that Bhāskara II was something of a poet, as were many Indian mathe-

Bhāskara II

Bhāskara II

maticians of the time who often wrote their work in the meter (rhythm) of a poem. Scholars also know that Bhāskara II had a son, Loksamudra, to whom he passed on his mathematical knowledge. It has also been written that Bhāskara II was a noted astrologer (one who believes that the stars and planets influence human affairs), and a story is told that he predicted that there was only one perfect moment for his daughter to be married. He had calculated exactly when that time would be, and arranged it so that a cup with a tiny hole in it would sink in a vessel of water at that proper time. When his daughter unknowingly allowed a pearl to fall in the cup and block the hole so that it did not sink, Bhāskara II refused to ever allow his daughter to marry once that magic moment had passed. Another version of this story says that his daughter's mistake made her wedding take place at the wrong time, and she lost her husband soon after they were married. These bittersweet stories may or may not be true, but both say that to make it up to his daughter or to console her, he wrote and dedicated the *Līlāvatī* to her. Scholars do know for a fact, however, that Bhāskara II was the best mathematician of his time and one whose work predated much of what Western mathematics was to achieve centuries later.

For More Information

"Bhaskaracharya II (1114–1185)." Simon Fraser University. http://www.math.sfu.ca/histmath/India/12thCenturyAD/Bhaskara.html (accessed July 8, 2002).

Narins, Brigham, ed. *World of Mathematics*. Detroit: Gale Group, 2001, p. 63.

The New Encyclopaedia Britannica Micropaedia. Vol. 2. Chicago: Encyclopaedia Britannica, 1998, pp. 186–87.

O'Connor, J. J., and E. F. Robertson. "Bhaskara." *The MacTutor History of Mathematics Archive*. School of Mathematics and Statistics, University of St. Andrews, Scotland. http://www.groups.dcs.st-andrews.ac.uk/~history/Mathematicians/Bhaskara_II.html (accessed July 8, 2002).

Pingree, David. "Bhāskara II." *Biographical Dictionary of Mathematicians*. New York: Charles Scribner's Sons, 1991, pp. 248–54.

Born 1832
Warwickshire, Gloucestershire, England

Died 1916
England

English mathematical learning theorist

Mary Everest Boole

For much of her long life, Mary Everest Boole was concerned with some aspect of teaching mathematics to children. As a teacher, lecturer, and writer, she had a farsighted vision of what the proper mathematical education should be like, and she always stressed the importance of communication and real understanding in learning mathematics. Her philosophy of education had a great impact on the progressive schools in England and the United States at the beginning of the twentieth century, and are useful even in modern day.

Born into a famous family

As the oldest child of Reverend Thomas Roupell Everest and Mary Ryall, Mary Everest and her younger brother, George, grew up in a privileged and an intellectually stimulating environment. Her father was an educated clergyman whose brother, English engineer George Everest (1790–1866), was surveyor general of India from 1830 to 1843, and the man after whom the world's highest point is named (Mount Everest, in the Himalayas between Nepal and Tibet). Another uncle, her mother's brother, John Ryall, was a professor of Greek and the vice president of Queen's College in Cork,

Ireland. Her uncle, George Everest, would make the family name famous with his mountain climbing, and he was very close to young Mary, to the point where he wanted to adopt her. However, it was through her other uncle, Ryall, that the young girl would meet some of the great English scientists of the time, such as English astronomer and mathematician John Herschel (1792–1871) and English mathematician Charles Babbage (1792–1871).

Mary Everest's introduction to mathematics came from her studies with a French tutor. When she was five years old, her father moved the family to Poissy (pronounced pwa-SEE), France, to be near his German doctor, Samuel Hahnemann (1755–1843). Hahnemann was the founder of homeopathy (pronounced hoe-mee-AH-peh-thee), a medical system that treated disease by giving the patient tiny doses of a drug that would cause symptoms of that disease in a healthy person. While in France, her father hired a tutor with the last name of Deplace for his five-year-old daughter, and this teacher would have a great influence over her. She became very fond of Deplace and his teaching style, which she never forgot. She later described it as leading children to new concepts by asking them a series of questions. The teacher would then have the students write their answers down immediately, and they would all analyze the questions and responses. This way, students would proceed on their own, with some guidance, to the right answers, and he believed that this was far superior to simply telling the children what the answers were. This way, their notebooks were filled with their own back-and-forth thoughts. This method also taught students to always ask questions and get answers.

Deplace's philosophy and methods had a great effect on Mary Everest, and she would later call him her hero. No doubt, her tutor must have enjoyed having a brilliant student who was fluent in two languages and whose father had introduced her to the **geometry** (see entry in volume 1) of **Euclid of Alexandria** (c. 325–c. 270 B.C.E.; see entry in volume 1) at the age of seven. At eleven, she was reading the advanced *Bonnycastle's Algebra* textbook on her own.

Meets George Boole

Mary and her family returned to England when she was eleven, and although she was no longer with her beloved tutor, she had access to her father's library and, more importantly, the stimulating circle of great minds she had earlier left behind. It was in this social circle that she first met English algebraist and logician **George Boole** (1815–1864; see entry in volume 1). In 1844, the Royal Society of London had awarded him a gold medal for a highly original paper discussing how **algebra** (see entry in volume 1) and calculus could be combined. When Mary first met him, she was only eighteen and he was thirty-five. He became her tutor, and five years later, her husband.

During the next nine years, the Booles had five daughters, and Mary Everest Boole helped her husband in his work. As one of the few people capable of understanding his original work, she acted as his secretary, editor, and even coworker at times. In fact, it was only with her great help that he managed to complete his pioneering work, *An Investigation of the Laws of Thought,* in 1854, just before they were married. This landmark work founded what is called mathematical **logic** (see entry in volume 2), or symbolic logic.

Mary Everest Boole was widowed in 1864 when her husband suddenly died of pneumonia. It is said that she may have made his illness worse or even contributed to his death by choosing an unusual treatment. George Boole had become sick after being caught in a cold rain, so she followed Hahnemann's "law of similars" and reasoned that whatever caused his illness would also provide the cure. She therefore put him to bed and then dumped buckets of water over him. After her husband died, Boole took a job as a librarian at Queen's College, which was England's first college for women. In fact, she was really more of an unofficial teacher and advisor to many of the students, and here she realized how much she loved teaching.

Teaches and writes about education

More and more, Boole found herself thinking about how children learn and how they are taught, especially about mathematics. She then took to writing some of her theories, and wrote her first book, *The Message of Psychic Science to Mothers and Nurses.* However, the controversy surrounding the subject of the spirit world got her fired from Queen's College. She then opened her home to guests and conducted her "Sunday night conversations" in which she discussed many of the scientific topics of the day with students. At the age of fifty, she began to write a series of books and

Mary Everest Boole

articles, and concerned herself more with the psychological aspects of teaching and education. She came to think of herself as a mathematical psychologist, and viewed students as individuals who used their minds, bodies, and even their unconscious—their whole selves—when they learned mathematics. She also did not believe that a competitive environment was necessarily a good way to learn at an early age. As noted in *Zimath,* Boole argued that "only dead mathematics can be taught where the attitude of competition prevails; living mathematics must always be a communal possession." In 1904, her book *The Preparation of the Child for Science* was published and, along with her *Lectures on the Logic of Arithmetic,* had a considerable influence on the progressive school movements of the early twentieth century.

Although widowed for fifty years, Boole supported herself, taught generations of children, wrote several books on mathematics and education, and raised five daughters—Mary, Margaret, Alicia, Lucy, and Ethel. One daughter, Alicia Boole Stott, followed in her parents' footsteps and became a mathematician; she was well known for her work on the visualization of geometric figures in hyperspace (space of more than three dimensions). Mary Everest Boole died in 1916 at the age of eighty-four. Her *Collected Works,* totaling more than fifteen hundred pages, were published in four volumes.

For More Information

Boole, Mary Everest. *A Boolean Anthology: Selected Writings of Mary Boole on Mathematical Education.* Compiled by D. G. Tahta. Derby, England: Association of Teachers of Mathematics, 1972.

Boole, Mary Everest. *The Preparation of the Child for Science.* Oxford: Clarendon Press, 1904. Reprint, International Society for General Semantics, 1978.

Cobham, E. M. *Mary Everest Boole: A Memoir with Some Letters.* Ashington, England: C. W. Daniel Co., 1951.

Frost, Michelle. "Mary Everest Boole." *Biographies of Women Mathematicians.* http://www.agnesscott.edu/lriddle/women/boole.htm (accessed July 9, 2002).

MacHale, Desmond. *George Boole: His Life and Work.* Dublin: Boole Press, 1985.

O'Connor, J. J., and E. F. Robertson. "George Boole." *The Mac-Tutor History of Mathematics Archive.* School of Mathematics and Statistics, University of St. Andrews, Scotland. http://www.groups.dcs.st-andrews.ac.uk/~history/Mathematicians/Boole.html (accessed July 9, 2002).

Surendran, Dino. "Mary Boole." *Zimaths.* http://www.uz.ac.zw/science/maths/zimaths/marybool.htm (accessed July 9, 2002).

Mary Everest Boole

**Born January 7, 1871
Saint-Affrique, France**

**Died February 3, 1956
Paris, France**

French number theorist

Émile Borel

*Émile Borel.
Reproduced by permission of
Mathematisches Forschungsinstitut
Oberwolfach.*

$\acute{\text{E}}$mile Borel was a major figure in twentieth-century mathematics whose pioneering work opened entire new fields of mathematics for others to study. Brilliant as a child, Borel led a long life in which he engaged in an enormous amount of mathematical research as well as involved himself with several social and political issues.

A genius and an achiever

Félix-Édouard-Justin-Émile Borel (pronounced baw-REHL) was born, the youngest of three children, in Saint-Affrique in the Aveyron (pronounced ah-vay-ROHN) district in southern France. His father, Honoré Borel, was a Protestant pastor in his village, and his mother, Emilie Teissie-Solier, was the daughter of a fairly wealthy businessman. Young Borel was educated at home by his father, and by the time he was eleven years old, everyone recognized that he had an extraordinary natural ability in mathematics.

In 1882, Borel's parents sent him to the nearest *lycée* (high school) which was at Montauban (pronounced maunt-oh-BAUHN), the nearest cathedral town. The educators immediately realized Borel had enormous potential and sent him off to Paris to study at some of the leading preparatory schools. Young Borel was able to join the family of Jean-Gaston Darboux (1842–1917), a mathematician in Paris whose son was his friend, and he lived as a member of the family. At school, he continued to demonstrate that he was a remarkable student, and in 1889, he entered a competition for admission to the leading engineering school (École Polytechnique), and the leading science and mathematics school (École Normale Supérieure). Borel was awarded first place for both exams, and had his choice of schools (both of which very much wanted him to attend). He eventually selected the École Normale Supérieure, the mathematical school and the best scientific institution in the country, and with which he would remain associated all of his career.

Begins his mathematical career

In choosing the École Normale, Borel deliberately selected what would be the life of a mathematician—research and teaching— over the considerably more practical world of engineering or technology. Little did he know, however, that his relationship with the École Normale would be such that fifty years later his colleagues would be celebrating the day he first joined that school. During his first year there, he excelled, publishing two papers. It was at the École Normale that Borel developed the lifelong habits that would make him so successful and so different from others. Borel was not only a brilliant individual, he was a person who was highly focused and could be intensely serious. To these traits he added incredible self-discipline, an ability to maintain the highest level of activity, and above all, an extremely high degree of organization. Further, although he had many good friends, he seldom wasted time with what people today call "small talk" or formalities, though he welcomed those who had something constructive to say.

To no one's surprise, Borel graduated first in his class in 1893 and was promptly offered a teaching position at the University of Lille. Such an offer to a new graduate was unheard of in its time, since it was most unusual for a French university to award a position to someone before receiving a doctorate degree. Borel accepted the Lille position, and remained there for three years, during which time he received his doctorate in 1894 from the École Normale Superieure. In 1896, he rejoined his favorite school and began a full career there in which he would redefine the way mathematics was taught in French universities.

Émile Borel

Émile Borel

In 1901, Borel married Marguerite Appell, the daughter of French mathematician Paul Appell (1855–1930). Borel had known her for some time, but he had to wait until she was seventeen years old before they could marry. His wife became a very productive novelist, writing over thirty books under the name Camille Marbo, and she was always very supportive of her husband's work. The couple had no children, although they did adopt one of Borel's nephews.

A diverse and productive mathematician

Borel began his career of research by publishing three brief articles at the age of eighteen. From there, he went on to a professional life in which writing and publishing would play a major role. This is seen in his over three hundred published papers and books. Borel worked in many mathematical fields, especially as a young man. One of the mathematical areas in which he will always be remembered for is his study of what are called complex functions. In mathematics, functions serve as a type of formula in which different numbers, known as variables, are plugged into a mathematical expression for a certain value. Borel was able to use his work on this to prove a theorem (a formula or a proposition) that had baffled other mathematicians for twenty years.

Borel also is recognized as having founded the field known as measure theory. This is defined as the study of lengths and **volume and surface area** (see entry in volume 2) in general spaces. Measure theory has recently become useful in the modern study of fractals. Fractal geometry is a **geometry** (see entry in volume 1) that gives modern science a way to deal with the randomness and uncertainties of nature. It has proven critical to understanding such natural phenomena as weather forecasting and the distribution of earthquakes. It was Borel's 1896 book, *Leçons sur la théorie des fonctions* (Lessons on the Theory of Functions) that laid the foundation of measure theory and became a classic.

Few seem to recognize that Borel's brilliant work on what is now called game theory actually was done a few years before Hungarian American mathematician and physicist **John von Neumann** (1903–1957; see entry in volume 2) began investigating this new field in 1926. In a series of papers begun in 1921, Borel was the first to define games in terms of strategy and to consider the different types of strategies there could be. He even applied them to war

Émile Borel

Hungarian American mathematician and physicist John von Neumann, whose work on game theory came a few years after that of Émile Borel. *Reproduced by permission of the Corbis Corporation.*

and economics, attempting to predict outcomes, which von Neumann and others would do some years later. His work in this new field was either ignored or not understood, and many consider Borel and not von Neumann to be the founder of game theory.

An involved life in politics

Borel had another side to his life aside from his work in mathematics. He was interested in and cared strongly about social and

Émile Borel

political issues of his day, and tried to do something about them. As early as 1906, he used prize money he won to start a monthly journal that considered social and cultural issues. During World War I (1914–18), he served his country by researching the science of sound location and helping to direct France's military research and development. Although he was a resident of Paris, he was elected mayor of his hometown of Saint-Affrique, and in 1925, he was appointed minister of the navy.

Many consider one of Borel's most lasting accomplishments to be his work helping to establish both the National Center for Scientific Research and the Henri Poincare Institute. Both still exist and owe their existence to Borel's belief that governments should support mathematics and the sciences. During World War II (1939–45), Borel played an even greater role after he was captured and detained for part of a year by the conquering German Nazis. Once released, he served with distinction in a group known as the Resistance, an underground organization in France that worked to free France from German rule.

Borel's accomplishments, awards, honors, and titles are enough for many men, and he received such recognition both during and after his life. He was honored as a statesman and civil servant as well as for his astounding mathematical achievements. A man who traveled extensively, he died in Paris after falling on board a ship returning from a conference in Brazil. He was mourned by mathematicians and nonmathematicians alike.

For More Information

Abbott, David, ed. *Mathematicians.* New York: Peter Bedrick Books, 1985.

Franceschetti, Donald R., ed. *Biographical Encyclopedia of Mathematicians.* New York: Marshall Cavendish, 1999.

May, Kenneth O. "Émile Borel." In *Biographical Dictionary of Mathematicians.* New York: Charles Scribner's Sons, 1991, pp. 315–18.

O'Connor, J. J., and E. F. Robertson. "Émile Borel." *The MacTutor History of Mathematics Archive.* School of Mathematics and Statistics, University of St. Andrews, Scotland. http://www.groups.dcs.st-andrews.ac.uk/~history/Mathematicians/Borel.html (accessed July 9, 2002).

Born c. 1455
Paris, France

Died c. 1500
Lyons, France

French arithmetician and algebraist

Nicolas Chuquet

Nicolas Chuquet is an example of a mathematician far ahead of his time. Working with the mathematical tools and concepts of the Middle Ages (the period of time in Europe between 500 and 1450), Chuquet offered a clear explanation of the role of zero and regularly used symbols. He also anticipated the invention of **logarithms** (see entry in volume 2) and was the first to use a radical sign with an index to indicate roots.

Few details of his life

Very little is known about the life of French mathematician Nicolas Chuquet (pronounced nih-ko-LAH shue-KAY). Many of the known facts about his life come from what Chuquet tells about himself in the one manuscript he wrote. In this book, Chuquet calls himself a Parisian, so it is assumed he was born in Paris. He also tells his readers that he held a bachelor's degree in medicine from a school in Lyons, a city in central France. After spending his youth in Paris, he moved to Lyons around 1480. Historians know this because his name appears in the Lyons tax register at that time. He may have practiced medicine in Lyons, but there is no evidence of that. However, he was certainly a well-educated

Nicolas Chuquet

person if he went to school long enough to obtain a bachelor's degree in medicine.

By 1485, Chuquet appears in the Lyons register as "Nicolas Chuquet, algoriste" (master of algorithms). An algorithm (pronounced AL-guh-rith-um) is any systematic method of doing mathematics that involves a step-by-step procedure. Historians have read his one work closely, trying to find clues that would tell them more about the man and what influenced him. For example, there is evidence of some Italian influence in his writings. Some think this means he may have visited Italy, while others say it exists only because there was a large Italian community living in Lyons at the time Chuquet was there.

An outrageous copycat

Chuquet titled the one book he wrote *Triparty en la science des nombres* (Three Parts in the Science of Numbers). Although this work is now recognized as being the earliest French thesis on **algebra** (see entry in volume 1), it was not published in any form during his lifetime. However, in 1520, a native of Lyons named Estienne de La Roche (1470–1530) published a book on arithmetic and algebra. Although he mentions Chuquet's *Triparty* at the beginning of his book, he did not state that he copied Chuquet's work, word-for-word. This was confirmed centuries later when Chuquet's original manuscript of *Triparty* was found and contained notes in La Roche's handwriting. When this was finally discovered in 1841, mathematical history was rewritten and all of the praise for La Roche's work was transferred to Chuquet.

Shows he is ahead of his time

Today, historians know that Chuquet was a man whose work shows that he possessed extensive learning. His writing demonstrates that he had a solid historical foundation, as he discusses the contributions of such great Greek mathematicians as **Euclid of Alexandria** (c. 325–c. 270 B.C.E.; see entry in volume 1), **Archimedes of Syracuse** (287–212 B.C.E.; see entry in volume 1), and **Claudius Ptolemy** (c. 100–c. 170). Chuquet's language was always simple and direct, and he did not boast or claim that everything he wrote was original to him. Examination of his text, however, shows that he was far ahead of most of his mathematical

Luca Pacioli

Luca Pacioli.
*Reproduced by permission of the
Corbis Corporation.*

It is believed that Nicolas Chuquet visited Italy at some time, and if that were so, he may have actually met the Franciscan monk named Luca Pacioli (1445–1517). Today, Pacioli is best known for having written what was the earliest printed book on arithmetic and algebra. Published in 1494 in Venice, Italy, it was titled *Somma di arithmetica, geometria, proporzione e proporzionalita* (The Whole of Arithmetic, Geometry, Proportion, and Proportionality). Since printing with movable type had only recently been invented by German Johann Gutenberg (c. 1398–c. 1468), Pacioli's book is among the first generation of modern or mechanically printed books.

Pacioli's work linked mathematics with a variety of practical applications. He is regarded as one of the first to introduce double-entry bookkeeping to the West. Also called the "method of Venice," this is a way of recording business transactions that allows a person to keep accurate records since it divides entries into debits (on the left side of a ledger) and credits (on the right). This simple method kept track of money coming in and going out.

Pacioli was somewhat of a wandering teacher throughout Renaissance Italy, and when he became the first occupant of a chair of mathematics founded by the Sforza family in Milan, he also met and became a close friend of the great Italian artist Leonardo da Vinci (1452–1519). Pacioli taught Leonardo mathematics and Leonardo helped illustrate one of Pacioli's books. Pacioli's book was important in its time because it contained much more than was taught in universities, and emphasized how mathematics could be used in the practical world of commerce. Pacioli was such good friends with Leonardo, that when the French army entered the city of Milan, they fled together for the safety of Florence.

counterparts. For example, unlike nearly everyone of his time, he used his own mathematical symbols, which he based on the appropriate French terms for the functions. For **addition** (see entry in volume 1), he would use a lower case *p* with a line above it, which stood for the French word *plus* meaning "more." For **subtraction** (see entry in volume 2), he used a lower case *m* with a line above it, taken from the French word *moins* for "less."

Chuquet also was the first to use a radical sign with an index to indicate roots. The symbol for a root $\sqrt{}$ is called a radical sign.

Nicolas Chuquet

The word radical comes from the Latin word *radix,* meaning "root." The number inside and on the right of the radical sign is called the radicand. The number in the upper left-hand corner of the radical sign tells which root is to be extracted and is called the index. It is usually written in a smaller type size than the radicand. In $\sqrt[3]{8},$ 8 is the radicand and 3 is the index.

Some of Chuquet's other accomplishments include his regular use of **fractions** (see entry in volume 1), which he called *nombres routz,* or broken numbers. He also used both positive and negative numbers, and gave rules for working with them. Chuquet played an important role in convincing Western or European mathematicians that they should incorporate the notion of zero into their thinking and their mathematics. Ever since the idea of using zero as a place indicator and as a number itself was introduced from the East by Italian number theorist Leonardo Pisano Fibonacci (c. 70–c. 1240), there had been considerable resistance to using zero. This Western reluctance to use zero dates back to Roman times, and no one can satisfactorily explain why it existed. However, as mathematicians like Chuquet wrote of the advantages of using zero, and explained clearly why this was so, this traditional stubbornness was slowly broken down.

Finally, Chuquet made the beginnings toward a discovery of logarithms. Logarithms are numbers known as **exponents** (see entry in volume 1) that are used to express the repeated **multiplication** (see entry in volume 2) of a single number. Chuquet was aware that 2^2 equals 4 ($2 \times 2 = 4$), and that 2^3 equals 8 ($2 \times 2 \times 2 = 8$), and he actually created a short table from 0 to 20 that worked out all the powers of 2. This was a century or more before the idea for logarithms was discovered and fully calculated. Today, Chuquet's original manuscript is in the Bibliothèque Nationale, the national library of France. It was not printed on its own or as a book until 1880, but by then, everyone recognized that this highly original work belonged to Chuquet and was not the work of La Roche.

For More Information

Flegg, Graham, Cynthia Hay, and Barbara Moss, eds. *Nicolas Chuquet, Renaissance Mathematician: A Study with Extensive Translation of Chuquet's Mathematical Manuscript Completed in 1484.* Boston: D. Reidel Pub. Co., 1985.

Itard, Jean. "Nicolas Chuquet." *Biographical Dictionary of Mathematicians.* New York: Charles Scribner's Sons, 1991, pp. 497–503.

O'Connor, J. J., and E. F. Robertson. "Luca Pacioli." *The MacTutor History of Mathematics Archive.* School of Mathematics and Statistics, University of St. Andrews, Scotland. http://www.groups.dcs.st-andrews.ac.uk/~history/Mathematicians/Pacioli.html (accessed July 9, 2002).

O'Connor, J. J., and E. F. Robertson. "Nicolas Chuquet." *The MacTutor History of Mathematics Archive.* School of Mathematics and Statistics, University of St. Andrews, Scotland. http://www.groups.dcs.st-andrews.ac.uk/~history/Mathematicians/Chuquet.html (accessed July 9, 2002).

Nicolas Chuquet

Born January 8, 1888
Lublinitz, Upper Silesia (now Poland)

Died January 27, 1972
New Rochelle, New York

German American algebraic geometer

Richard Courant

Richard Courant.
Reproduced by permission of
AP/Wide World Photos.

Richard Courant specialized in the application of mathematics to physics. He contributed significantly to the rebirth of applied mathematics in the twentieth century by founding two highly influential mathematical institutes.

Independent-minded and self-supporting

Richard Courant (pronounced COOR-awnt) often said that he had supported himself from the age of fourteen, and indeed he did. This came about because of his family's financial problems and his own independent spirit. The first of three sons, Courant was born in Lublinitz, a small town in Upper Silesia, which was then in Germany but is now part of Poland. His father, Siegmund Courant, had his own business, and his mother's name was Martha Freund. The family moved to Glantz when Courant was three, and when he was nine, they moved to Breslau (now Wrocław), the capital of Silesia. There, Courant attended the Konig Wilhelm Gymnasium, a preparatory school for college. Although he struggled there at first, he eventually became an excellent student. In 1902, his father's business went bankrupt, forcing the family to move again, this time to Berlin. Young Courant, however, stayed behind in Breslau to attend school.

By the time he was fourteen, Courant decided that he would help his family with their financial problems by earning some money himself. After answering a newspaper advertisement for a tutor and then getting the job, he soon built up his own business tutoring other young people in mathematics. By 1905, his last year in high school, he stopped going to class since he was earning so much money. He made up for his lost high school time by attending lectures in mathematics and physics at the local university. Oddly enough, Courant found himself tutoring students to take the high school mathematics final exam, which he himself had not yet taken. Because of this and other irregularities, his school eventually asked him to leave. The following spring, Courant took and passed that final and then immediately took and passed the university entrance examination. In 1906, he enrolled in the University of Breslau to study physics.

From physics to mathematics

It did not take Courant very long, however, to realize how poor the physics instruction at Breslau was, and after switching to mathematics, he entered the University of Zürich. After one semester there, he left and transferred to the University of Göttingen to take courses with German number theorist David Hilbert (1862–1943). Although Courant still thought he would become a physicist, he found the physicists at Göttingen to be nowhere near as inspiring and exciting as the mathematicians. Courant made the final break from physics and fully embraced mathematics after he became an assistant to Hilbert and chose a thesis topic under his direction. He obtained his Ph.D. in mathematics from Göttingen in 1910.

Marriage and wartime

After spending 1911 doing a year of required military service in the German army, he returned to Göttingen as an unsalaried lecturer. The next year he became salaried (was paid regularly to teach). This allowed him to marry, and in 1912 he married Nelly Neumann, a young woman he had tutored for many years. They had no children and were divorced four years later.

Richard Courant

During these years, World War I (1914–18) broke out and Courant received his order to join the German army in 1914. He served as an infantryman until he was wounded in 1915, and spent the rest of the war working for the wireless communication (telegraph) department. There, he suggested many improvements, and since his job was to train soldiers how to use this equipment, he probably was spared from being killed. When the war ended late in 1918, he returned to Göttingen where he resumed teaching. Having gotten a divorce during the war, he was free to marry Nerina Runge, the daughter of German mathematician and physicist Carl D. T. Runge (1856–1927). The couple married on January 22, 1919, and would eventually have four children.

Makes Göttingen an international center

By 1920, Courant had become a professor at Göttingen, and during the next five years he would succeed in making Göttingen an international center of theoretical and applied mathematics. In 1922, when Courant founded Göttingen's Mathematics Institute, it was only an idea and did not even have its own building. However, using his remarkable skills as a teacher, researcher, and above all organizer and administrator, Courant succeeded in making his Institute the world's leading center for mathematical research. He did this partly by making no distinction between pure or theoretical and applied mathematics, and therefore was able to attract physicists from all over the world. Göttingen and the Institute soon became the center for research in the new field of quantum mechanics. During the late 1920s, Courant worked with Hilbert on his most important publication, *Methoden der mathematischen Physik* (Methods of Mathematical Physics). This important work was tremendously successful since it laid out the basic mathematical techniques that would play a role in the new quantum theory and in nuclear physics.

New York years

While Courant and his Institute were prospering, Germany was experiencing the rise of dictator Adolf Hitler (1889–1945), whose racist policies would eventually result in the deaths of some six million Jewish people. In 1933, when Hitler's Nazi Party came to power, Courant was expelled from Göttingen because he was Jewish. This move reflected the National Socialist (Nazi) government's campaign against all Jews as well as its unhappiness with the Göttingen Institute, which it believed was a center of dangerous independent and liberal thought. After accepting an invitation to teach at Cambridge University in England, Courant left Germany in 1933. The next year, he accepted an invitation to join the teaching staff at New York University, and he moved to the United States.

In 1936, after he was appointed head of New York University's department of mathematics, he did for New York University what he had done for Göttingen by creating a center of mathematics and science that would become internationally important. This was a remarkable accomplishment since, unlike Göttingen, which already had a group of highly talented individuals, at New York he had to start from scratch. Nonetheless, his ability to attract promising young mathematicians, and his willingness to teach them and help them get published and to get financial support was a great part of his successful organizing skills.

During World War II (1939–45), Courant was a member of the Applied Mathematics Panel, which assisted scientists involved with military projects. His post-war work in mathematics also played a vital role in the development of computer applications to scientific work. In 1940, he became an American citizen and cowrote a highly successful book called *What Is Mathematics?* which explained his subject to the public in nontechnical terms.

In 1953, Courant's applied mathematics research center at New York University became the Institute of Mathematical Sciences with him as its director. He stayed in that position until he retired in 1958 at the age of 70. In 1964, New York University changed the name to the Courant Institute of Mathematical Sciences to honor his great efforts and achievements. During his lifetime, he was honored with the U.S. Navy Distinguished Public Service Award as well as the Knight-Commander's Cross and the Star of the Order of Merit of the Federal Republic of Germany. In 1965, he received an award for distinguished service to mathematics from the Mathematical Association of America. Courant suffered a stroke on November 19, 1971, and died two months later. At a memorial in his honor, it was said that his work "never stood alone; it was always connected with problems and methods of other fields of science, drawing inspiration from them, and in turn inspiring them."

Richard Courant

Richard Courant

For More Information

Abbott, David, ed. *Mathematicians.* New York: Peter Bedrick Books, 1986.

Courant, Richard, and Herbert Robbins. *What Is Mathematics? An Elementary Approach to Ideas and Methods.* New York: Oxford University Press, 1941. Rev. ed., 1996.

Mathematicians and Computer Wizards. Detroit: Macmillan Reference USA, 2001.

O'Connor, J. J., and E. F. Robertson. "Richard Courant." *The MacTutor History of Mathematics Archive.* School of Mathematics and Statistics, University of St. Andrews, Scotland. http://www.groups.dcs.st-andrews.ac.uk/~history/ Mathematicians/Courant.html (accessed July 9, 2002).

Born July 31, 1704
Geneva, Switzerland

Died January 4, 1752
Bagnols, France

Swiss geometer and probability theorist

Gabriel Cramer

Gabriel Cramer.
Reproduced by permission of the
Granger Collection.

Gabriel Cramer was an influential teacher and editor who also introduced the concept of utility to mathematics. His mathematical work covered a wide range of subjects, and he later specialized in the history and the philosophy of mathematics. With much of his work overshadowed by people of his generation who are more famous, he is best known as someone who helped circulate the mathematical ideas of others.

Impressive as a youth

Gabriel Cramer was born in Geneva, Switzerland, on July 31, 1704. The century before, his family had moved south from northern Germany to Strasbourg, and eventually to Geneva where they remained. His father, Jean Isaac Cramer, was a doctor, and his mother's name was Anne Mallet. He had two brothers, Jean-Antoine, who also became a doctor, and Jean, who became a professor of law. Historians do not know exactly where young Cramer attended school, except that it was in Geneva and that he showed early promise in science. He must have been a remarkable student since scholars know that he defended his thesis on the theory of sound in 1722 at the age of eighteen. This means that he was awarded a doctorate while still a teenager. Not surprisingly, two years later, he was in a competition for the chair of philosophy at a Geneva university.

Gabriel Cramer

Becomes cochair

The competition took place in 1724 when the Académie de Calvin in Geneva held a contest for its chair of philosophy. There were only three contestants, one of whom was Cramer. The others were Amédée de la Rive and Giovanni Ludovico Calandrini (1703–1758). Calandrini was only a year older than Cramer, and de la Rive was the oldest of the three. Although de la Rive eventually was given the position because he had the most experience, the committee was very impressed by the other two young candidates. Because the committee members considered the futures of Calandrini and Cramer to be so bright, they came up with a way of retaining them as well.

The Académie committee decided to split the chair of philosophy into two chairs, one chair of philosophy and one chair of mathematics. De la Rive was given the chair of philosophy that he had won, and the other new chair was offered to both Cramer and Calandrini on the condition that they would share it. As cochairs, only one of them would occupy the position at any one time. They would alternate teaching full time, therefore splitting the duties and the salary. In addition, while one was busy teaching, the other man would travel for two or three years in Europe, meeting and learning from other renowned mathematicians.

This unusual situation suited both Cramer and Calandrini, and they became good friends. The two also divided up the courses they taught, with Cramer teaching **geometry** (see entry in volume 1) and mechanics or physics, and Calandrini teaching **algebra** (see entry in volume 1) and astronomy. As professors, the two broke with tradition and taught their classes in French rather than in Latin. Latin was the traditional language of science and of all scholars, but the two men realized that they would reach a larger audience by using French. As noted in *The MacTutor History of Mathematics Archive*, Cramer believed this teaching method would be especially helpful to those "persons who had a taste for these sciences but no Latin."

Travels and forms lasting relationships

After teaching at the Académie since his appointment in 1724, Cramer eagerly set out in 1727 for two years of traveling. As an intel-

Gabriel Cramer

Johann Bernoulli, whose works were frequently edited by Gabriel Cramer.
Reproduced by permission of the Corbis Corporation.

ligent, healthy, pleasant twenty-three-year-old with good references, he was welcomed by the leading mathematicians of his time. These two years proved to be an important time in his professional life. First, he went to Basel, Switzerland, and spent five months with Swiss mathematician Johann Bernoulli (1667–1748), best known for his pioneering efforts in calculus. There, he also met Swiss geometer, physicist, and number theorist **Leonhard Euler** (1707–1783; see entry in volume 1), who was considered the premier mathematician of that century. Before returning to Geneva in 1729, Cramer visited

England, the Netherlands, and France, and met with all of the top mathematicians and scientists in those countries. Besides learning a great deal, he also formed many life-long friendships.

Gabriel Cramer

In 1729, Cramer returned to Geneva to resume teaching. In 1734, the "twins," as Cramer and Calandrini were called, finally split when de la Rive left the Académie and Calandrini was given the chair of philosophy. Cramer no longer had to share the chair of mathematics. Much later, in 1750, Cramer assumed the philosophy chair when Calandrini left to enter government service.

Mathematical work and editing

During his fairly short life, Cramer made a number of original contributions to mathematics, and one of the best-known is called Cramer's Rule. In 1750, he discovered a simple procedure for solving systems of simultaneous linear equations. Linear equations are used in algebra; the most basic form of algebra involves equations that are called linear or first-degree equations. They have this name because none of their terms are raised to a power of two or higher, and when they are represented in **coordinate graphing** (see entry in volume 1), they always result in a straight line.

Cramer also introduced the idea of utility to mathematics. This was a very sophisticated concept for its time, and today it is used to link **probability** (see entry in volume 2) theory with mathematical economics. Utility simply means that something has the ability to satisfy a human want or need. Utility theory states, naturally, that rational or reasonable people always try to maximize the expected utility of something. This theory has proved useful to modern game theory and theories of decision-making.

Besides his original work, Cramer's most lasting contribution to mathematics is the editing of other people's work. Often a thankless and almost endless job, Cramer brought intelligence, understanding, and tirelessness to it. It was here that his multiple personal connections to mathematicians throughout Europe really paid off, as he was able to perform the extremely valuable task of editing some of the best work of the best mathematicians. For example, a few years before he died, Bernoulli arranged for his friend Cramer to edit and publish his complete works. In doing this for Bernoulli, Cramer became the first real mathematical scholar to edit the work of another, meaning that he was on the same intellectual level of the person whose work he was arranging and editing. In fact, Bernoulli authorized that only Cramer should collect, edit, and publish his work, and that there should be no other edition.

Cramer edited other famous mathematicians' work as well. Bernoulli arranged for Cramer to edit the papers of his brother, Jakob Bernoulli (1654–1705), also a mathematician, who was best known for his work in probability. Cramer also edited and published the correspondence between Johann Bernoulli and German philosopher and logician (one who uses reason) **Gottfried Leibniz** (1646–1716; see entry in volume 2). Cramer also edited the five-volume work by German philosopher, mathematician, and scientist Christian von Wolff (1679–1754). Cramer's accomplishments as a mathematics editor were vital to the support, circulation, and preservation of mathematical knowledge.

Cramer never married and, like his two brothers, he always took an interest in local government. Throughout his career, he regularly served the city of Geneva on several major councils, and also was a consultant on its defenses and cathedral repairs. Always a healthy, although overworked, individual, he had a sudden end when he fell from a carriage and could not fully recover. As his health failed, his doctors advised that he travel to the warmer regions of southern France, and it was during this overland journey that he died at the age of forty-eight about twenty-five miles north of Avignon.

Gabriel Cramer

For More Information

Abbott, David, ed. *Mathematicians*. New York: Peter Bedrick Books, 1986.

Franceschetti, Donald R., ed. *Biographical Encyclopedia of Mathematicians*. New York: Marshall Cavendish, 1999.

Jones, Phillip S. "Gabriel Cramer." In *Biographical Dictionary of Mathematicians*. New York: Charles Scribner's Sons, 1991, pp. 545–49.

O'Connor, J. J., and E. F. Robertson. "Gabriel Cramer." *The MacTutor History of Mathematics Archive*. School of Mathematics and Statistics, University of St. Andrews, Scotland. http://www.groups.dcs.st-andrews.ac.uk/~history/Mathematicians/Cramer.html (accessed July 10, 2002).

Born 1824
Hamburg, Germany

Died 1861
Place unknown

German calculating prodigy

Zacharias Dase

Zacharias Dase may be the most phenomenal mental calculator who ever lived. Although he was only average and often below average at nearly everything else he attempted, Dase could quickly and correctly calculate numbers in his head that took others a long time to figure out with pencil and paper. Dase met some of the great mathematicians of his time and travelled throughout Europe giving exhibitions of his amazing memory and calculating ability.

"The lightning calculator"

Johann Martin Zacharias Dase (pronounced DAAS-zuh) was born in Hamburg, Germany, and is said to have received a fair education. He did not come from a wealthy or privileged family and, therefore, went to local public schools. Nothing is known of his parents or his family life. What is known, however, is that he was recognized as a real calculating prodigy (a child with some type of exceptional talent). Dase's ability to do nearly instant calculations in his head were so extraordinary that he eventually became known as "the lightning calculator." But Dase was not able to transfer or use his mental powers in any other field, nor was he even able to learn **geometry** (see entry in volume 1) or any other language but his native German.

Savant syndrome

In more modern times, Dase might be considered an "idiot savant" (pronounced EE-dee-oh suh-VAHN). An idiot savant is a person who has extraordinary skills in a very specialized area but who is, nonetheless, intellectually slow or mentally limited. This condition is now called the savant syndrome. Although overall a rare condition, the mathematical calculator is perhaps the most common form of savant syndrome. These mental calculators often display incredible calendar memory or can calculate square roots and prime numbers in their heads at amazing speeds. Others with savant syndrome are musical geniuses and can hear a complicated piece of music and immediately play it back perfectly from memory. Other persons are gifted artists.

One thing all these individuals share is that their special skill is always in an extremely deep but narrow range. Dase was this way, and was no different from other great mental calculators, some of whom lacked even basic arithmetic skills. For example, Dase was unable to learn geometry. In almost every case of savant syndrome, people have very rigid limits concerning what they are able to do with the talent they have. Although one might be able to recite all manner of detail or facts about something, or quickly multiply large numbers, he or she is not able to apply that information in any other way than the narrow method they are used to. While there have been some advances in understanding this syndrome, all attempts to completely explain it are still educated guesses. Most would admit that scientists are still baffled in trying to understand this fascinating phenomenon.

Legendary multiplying ability

During his short life, Dase performed some truly amazing mental arithmetic. For example, when he was taken before some of the great mathematicians of his time and given certain problems to perform, he awed even them. When asked to multiply 79,532,853 by 93,758,479, he answered correctly in fifty-four seconds. He then took only six mental minutes to find the correct product of two twenty-digit numbers. Another time, he extracted the square root of a hundred-digit number in fifty-two minutes. Dase was human, however, and he was known to make mistakes whenever he had a headache.

By the time Dase was sixteen years old, he was already performing or conducting public exhibitions of his mental calculating ability.

Zacharias Dase

Zacharias Dase

In 1840, Dase was introduced to several well-known German scientists, one of whom was legendary German mathematician and astronomer **Carl Friedrich Gauss** (1777–1855; see entry in volume 1). Gauss eventually arranged for the Hamburg Academy of Science to hire Dase so he could devote full–time attention to his calculating. In less than two months, Dase had computed **pi** (see entry in volume 2) to 200 **decimal** (see entry in volume 1) places all in his head, and his work was published in the prominent *Crelle's Journal* in 1844. Dase also calculated the natural **logarithms** (see entry in volume 2; the power of which a base of ten must be raised to produce a number) of the first 1,005,000 numbers to 7 decimal places, and did some calculating work for the Austrian government.

For most people with savant syndrome, their impressive feats are usually tied in some way to memory, and almost all have an extraordinary memory for detail. Dase was no exception, and he gave public exhibitions of his mental calculating powers in Germany, Austria, and England. He could repeat all of the numbers he used in one of his performances as long as an hour later. Further, he had the striking gift of stating the correct number of things in a group after only a single glance, since he was somehow able to visualize and recall a large number of objects. For example, he could immediately tell the number of sheep in a flock or books in a case after only a quick look. He could also glance at a row of dominoes and quickly total their points. At other times, he could instantly give the correct number of letters in a certain line of print chosen at random.

Aside from his unique mental powers, Dase was described as a sort of dull or even slow person who was considered always trustworthy. His last years were spent in factoring the numbers between seven million and ten million for the Hamburg Academy. Such tasks were especially important in those days because there were no calculators or computers. Dase had finished nearly half of his task for Hamburg when he died at the age of thirty-seven.

For More Information

Mallette, Vincent. "History and Oddities of the Number Pi." *Journal of the Tennessee Academy of Science* (vol. 37, no. 4, October 1962). http://www.inwit.com/inwit/writings/history andodditiesofpi.html (accessed July 10, 2002).

Newman, James R. *The World of Mathematics.* Vol. 1. Redmond, WA: Tempus Books, 1988, pp. 465–67.

O'Connor, J. J., and E. F. Robertson. "Johann Martin Zacharias Dase." *The MacTutor History of Mathematics Archive.* School of Mathematics and Statistics, University of St. Andrews, Scotland. http://www.groups.dcs.st-andrews.ac.uk/~history/Mathematicians/Dase.html (accessed July 10, 2002).

Preston, Richard. "The Mountains of Pi." *New Yorker* (March 2, 1992). http://www.barryland.com/pi.html (accessed July 10, 2002).

Zacharias Dase

Born February 21, 1591
Lyons, France

Died October 1661
France

French geometer and engineer

Girard Desargues

G irard Desargues was a highly original thinker whose ideas formed the basis for the development of projective **geometry** (see entry in volume 1). Although he accomplished a great deal as an engineer and architect, his mathematical work was either misunderstood or ignored during his lifetime, and its significance was not recognized until the nineteenth century.

Few details on early life

Although Girard Desargues (pronounced day-ZAHRG) came from a very well-to-do family, there are surprisingly few facts known about his beginning years. He was one of nine children born to Girard Desargues and Jeanne Croppet. The elder Desargues was a tax collector for the church and owned several large houses in Lyons (pronounced lee-OHN), France, as well as a chateau (a large country house) and several vineyards near that city. Both sides of Desargues's family had been very rich for several generations and had supplied France with many lawyers and judges. Although historians have no facts about his education, it is apparent that his family situation gave Desargues every opportunity to obtain the best education available.

The next thing known about Desargues is from a report that places him in Paris in 1626, where he was working on an engineering project on the Seine (pronounced SAYNE) River. In September of that year, he proposed to the Paris municipality that it should raise the level of the Seine using powerful machines in order to be able to pump water throughout the city. Two years later, historians know that Desargues served as an engineer during a blockade of La Rochelle. This Atlantic seaport was involved in the Protestant-Catholic conflict known as the Thirty Years' War (1618–48). When La Rochelle sided with the English (Protestants) against the French (Catholics) and King Louis XIII (1601–1643), the French built a huge sea wall (which is why they needed engineers like Desargues), which prevented the English from bringing in supplies. After fifteen months, three-quarters of the population had died of starvation and the town surrendered. It was during the blockade of La Rochelle that Desargues met French algebraist, geometer, and philosopher **René Descartes** (1596–1650; see entry in volume 1), who had made the military a part-time career. Desargues also is said to have worked as an engineer and technical advisor for the French government under the king's minister, Cardinal Richelieu (1585–1642; pronounced REE-shell-yeh).

Girard Desargues

Joins Paris intellectuals

By about 1630, Desargues was spending more time in Paris and eventually became acquainted with some of that city's major thinkers. After meeting some of the leading mathematicians, he became part of a choice intellectual circle around 1635 that included such figures as French mathematician Marin Mersenne (1588–1648; pronounced mur-SEHN), French philosopher Pierre Gassendi (1592–1655; pronounced gah-SAHN-dee), and Etienne Pascal (1588–1651) and his young son, **Blaise Pascal** (1623–1662; see entry in volume 3), who was a child with extraordinary talent and would become a major French mathematician and physicist. Desargues is said to have regularly attended the meetings of this group, known as the Académie Parisienne. Stimulated by these intellectuals, Desargues began to research, write, and publish. In 1636, he published a short work in which he laid out his plans to develop a universal set of geometrical principles based on what is known as perspective.

Girard Desargues

Perspective and geometry

Perspective is a drawing or painting technique discovered in the fifteenth century. It is a method of realistically capturing and transferring the three-dimensional world onto a flat, two-dimensional surface or canvas. Using perspective, the artist paints what he or she sees, so that objects in the distance appear to shrink, and parallel lines and planes always converge or meet as they approach the horizon. Using this technique, the artist can paint a realistic version of a certain scene by making it seem to have space and **volume** (see entry in volume 2). Desargues studied perspective and realized that Euclidean (pronounced yoo-CLUD-ee-an) geometry did not agree with this way of seeing the world. In Euclidean geometry, parallel lines never meet. Desargues knew that the painters who used perspective were not simply distorting things in random ways but were instead closely following a new and modern type of geometry that might be called the geometry of the eye. He therefore attempted to come up with a theory that would explain what he considered should be a new type of geometry. Desargues's theory would eventually come to be called projective geometry. It would get this name because it would study the properties involved in transferring or projecting a three-dimensional object to a two-dimensional canvas.

Projective geometry was not fully stated or understood until the nineteenth century when French mathematician Jean Victor Poncelet (1788–1867; pronounced pohns-LAY) turned it into a solid and accepted branch of mathematics. Once its principles became understood, it proved essential not only to drawing and painting but for many other practical purposes involving projection, such as maps of the Earth's surface, shadows cast by objects, and even motion pictures.

Desargues misunderstood and attacked

The problems that Desargues encountered in trying to get his new theorem across illustrate just how important the ability to communicate is to mathematics and all of science. In 1639, Desargues privately published a small book in which he attempted to describe his new ideas of projective geometry. It proved a disaster, however, since it was almost unreadable. For some reason, Desargues decided to use a highly original (and some say eccentric or even weird) vocabulary, and he gave things names that no one knew or recog-

Frank Lloyd Wright

Frank Lloyd Wright.
Reproduced by permission of the
Corbis Corporation.

As Girard Desargues was both a profoundly original mathematical thinker in the seventeenth century and an accomplished architect with a unique building style, he in many ways leads one to think of the greatest architect of the twentieth century, Frank Lloyd Wright (1867–1959). Besides architecture, both had a highly individualized sense of what geometry was all about. Where Desargues saw beyond the conventional, Euclidean geometry of his time toward a new form of geometry that used perspective, Wright felt that geometry was all about the human being and tried to design buildings that reflected purely geometric shapes.

As one of the founding fathers of modern architecture, Wright designed some of the world's most important architectural treasures. During his long life, he designed some 800 buildings, 380 of which were actually built and about 280 of which still exist. While he believed in "organic architecture," meaning that things should always be natural or fit their setting or environment, he also believed in geometry and its basic, everlasting traditional shape. Wright took some mechanical drawing and basic mathematics courses during his years at the University of Wisconsin, but he was by no means a mathematician. Despite this lack of formal mathematical training however, he had a natural sense and feeling for the geometry he found in nature, and he strove to duplicate this natural geometry in his designs.

Although Wright's architectural style was described as "Prairie style," it could vary from the sharp angles of his private homes to the circles and spirals of the Guggenheim Museum in New York City. Interestingly, Desargues was also well-known for his own version of elaborate spiraling staircases. For Wright, however, each design had to be unique since each had to relate to the person who was to live in it and the place in which it was set. In one of his more famous homes, the Frederick C. Robie House, in Chicago, Illinois, he used straight lines as a theme. There, parts of the house parallel its surrounding wall, and everything seems to have a horizontal feature about it. In his own home, Wright used the circle to create openness. In the Guggenheim Museum, he again used circles and spirals to bring the visitor up toward the natural skylight. In other homes, Wright's thematic use of hexagons, pentagons, octagons, and even half circles are clearly visible. Wright's love of geometry and his emphasis on the beauty of geometric shapes allowed him to incorporate a little bit of mathematics into every one of his designs.

nized. For example, he introduced as many as seventy new terms, many of which were based on botanical concepts, meaning he was borrowing and making up geometrical words that were more appropriate to the study of plants. This unclear and confusing style of writing may have been highly original, but it resulted first in its author being criticized and attacked, and eventually in the work itself being ignored until Poncelet rediscovered it in 1822. By 1645, Desargues had been involved in this bitter debate for some years, and he turned to architecture in order to express his creativity. He proved to be an accomplished architect as well as an excellent engineer, and designed several new homes in Paris as well as more houses and mansions in Lyons. He became particularly noted for his elaborate spiraling staircases.

Finally appreciated

The original and innovative mind of Desargues was not fully appreciated until his mathematical work was rediscovered over 150 years after his death. Despite his inability or unwillingness to make himself understood, he was a geometer of profoundly unique ideas who introduced the major concepts of projective geometry. He was undoubtedly the most original contributor to geometry in the seventeenth century. The revival and growth of geometry in the nineteenth century was due in part to the rediscovery of projective geometry, and today it is a popular subject because of its visual applications on computer screens.

It is not known exactly where or when Desargues died, although most believe he died in Lyons, France. A reading of his will in Lyons on October 8, 1661, said that Desargues had died several days before, although it did not specify the exact location, cause, and date of his death.

For More Information

Field, J. V., and J. J. Gray, eds. *The Geometrical Work of Girard Desargues*. New York: Springer-Verlag, 1987.

Franceschetti, Donald R., ed. *Biographical Encyclopedia of Mathematicians*. New York: Marshall Cavendish, 1999.

Frank Lloyd Wright Foundation. http://www.franklloydwright.org/ (accessed July 10, 2002).

Frank Lloyd Wright Preservation Trust. http://www.wrightplus.org/ index.html (accessed July 10, 2002).

McDonough, Yona Zeldis. *Frank Lloyd Wright.* New York: Chelsea House Publishers, 1992.

Middleton, Haydn. *Frank Lloyd Wright.* Chicago, IL: Heinemann Library, 2002.

O'Connor, J. J., and E. F. Robertson. "Girard Desargues." *The MacTutor History of Mathematics Archive.* School of Mathematics and Statistics, University of St. Andrews, Scotland. http://www.groups.dcs.st-andrews.ac.uk/~history/Mathematicians/Desargues.html (accessed July 10, 2002).

"Projective geometry." *Infinity.* University of Michigan. Department of Mathematics. http://www.math.lsa.umich.edu/~mathsch/courses/Infinity/Geometry/Lesson1.shtml (accessed July 10, 2002).

Taton, René. "Girard Desargues." In *Biographical Dictionary of Mathematicians.* New York: Charles Scribner's Sons, 1991, pp. 595–600.

Girard Desargues

Born June 17, 1898
Leeuwarden, Netherlands

Died March 27, 1972
Laren, Netherlands

Dutch graphic artist

M. C. Escher

Despite his lack of any formal training in mathematics, the master artist known as M. C. Escher created intricate and captivating drawings that so fascinated and stimulated mathematicians that they came to regard him as one of them. Considered a sort of intuitive (natural) mathematician, Escher had the rare ability to visually portray complex mathematical concepts with drawings that stunned viewers and left them puzzled, amazed, and even awestruck. His careful drawings of repeated geometric patterns and his studies of infinite space are regarded by some to be a form of mathematical research.

An early talent for drawing

Maurits Cornelis Escher (pronounced EH-skur) was born in 1898 in the Netherlands (also called Holland), a country on the North Sea in western Europe. He was the third and youngest son of a distinguished civil engineer, George Arnold Escher, and his second wife, Sarah Gleichman. In elementary school, Escher was not a very good student, and was especially bad in mathematics. His father did, however, recognize that the young boy had some artistic talent since he drew very well and very easily (and only got good grades in art). By the time he was a teenager, his artistic talent was obvious to all, and his family and friends encouraged him to become an architect. When his family moved to Arnhem in 1912 from Leeuwarden (pronounced LAY-var-den), a northern city near the sea but inland to Arnhem near the German border, he entered school there and began his serious study of art. At school in Arnhem, he was fortunate to have a teacher named F. W. van der Hagen who first taught him how to cut a piece of linoleum (a smooth floor covering made of a mixture of oils, gums, and resins) in order to make an engraving. In 1916, Escher completed his first graphic work, which was a linoleum cut of his father.

After moving to nearby Oosterbeck (pronounced OOZE-ter-bek) in 1917 with his family, he followed his father's advice and attended the Technical College in Delft for a year, and then entered the School for Architecture and Decorative Arts in 1919 in the city of Haarlem, which is close to the capital of Amsterdam. There, he met Samuel Jessurun de Mesquita, an instructor who would greatly influence his development as an artist. This friendly artist recognized the distinctive artistry and real genius of his young student and insisted that Escher give up his studies of architecture and instead concentrate on becoming what he really was—a unique and gifted graphic artist. Since Escher's father knew the young man's heart was not fully in architecture, he agreed to have Escher give it up and focus solely on his art.

Becomes an artist

Escher would remain at the Haarlem school perfecting his woodcut techniques until he graduated in 1922. During these years at school, he and his parents often took trips, and it was during a 1921 trip to Italy that he first became inspired by that country's cities and landscapes. Italy would remain a source of inspiration throughout his life. He continued to draw, and late in 1921 some of his woodcuts were used to illustrate a small book titled *Flor de Pascua* (Flower of Passover).

After graduating, Escher continued to travel throughout Europe, sometimes with two friends, and also continued to grow as an artist. Italy and Spain inspired him and, as he travelled, he seemed to draw everything he saw there—plants, landscapes, and even insects. From April to June 1922, Escher roamed northern Italy, and from September to November of the same year, he travelled

M. C. Escher

M. C. Escher

by freighter (a ship loaded with goods) to Spain. There, in the city of Granada (pronounced gruh-NAH-duh), he first saw examples of Moorish or Arabic decorative styles, which he studied and copied. This style would also have a lasting influence on his work. It was during this year that Escher sold his first print, titled *St. Francis (Preaching to the Birds)*. As noted in the *Escher Pages*, Escher claimed to have "worked like a madman" on this particular piece, which sold in large numbers. That same year, he did his first woodcut that had a mathematical basis, called *Eight Heads*.

Settles in Rome and becomes successful

After Spain, Escher returned to Italy and continued to work very hard on his art, producing woodcuts of landscapes that showed baffling and seemingly impossible perspectives. While he was living for an extended time at a hotel in Siena, a northern city near Florence, he met Jetta Umiker (pronounced YET-tuh UHM-ih-kur), a Swiss woman who was in Siena with her family for an extended stay. Over the next few months, Escher fell in love with her but was only able to tell her so just as the family was about to leave. Escher then continued to see Jetta. In August 1923, he proposed to Jetta and also had his first one-man show in Siena.

The couple married the next year and, after some travelling, settled in a new home in Frascati (pronounced frah-SKAH-tee), outside of Rome. Escher also had another one-man show in his own country and, by this time, his popularity and reputation were steadily growing. A good example of his fame is the fact that in July 1926, both Italian dictator Benito Mussolini (1883–1945), and Italian king Victor Emmanuel III (1869–1947) attended Escher's baby's christening. By 1929, his prints were selling very well and he had several more one-man shows. Although Escher loved Rome, he moved his family to Switzerland in 1935 when the Italian government turned completely oppressive and dictatorial. By then, Mussolini was leading Italy, which would eventually join forces with Germany during World War II (1939–45).

Escher and mathematics

One does not have to know anything about art, woodcuts, or even mathematics to know that there is something unique or at least very special about Escher's drawings. Nothing about them is commonplace or typical. His land-scapes, buildings, and even animals are all seen and drawn in a highly individualistic way. While his work was always different, his subject matter and style changed considerably after he left Italy. Revisiting Spain in 1936 apparently was the influence that brought about this dramatic switch.

Following his re-exposure to the intricate repeating patterns of the ancient Moorish or Arabic mosaics and tiles on the ceilings and walls of the Alhambra (pronounced ahl-HAM-bruh) palace, Escher began to make drawings of similar images that he had in his own mind rather than of the things he actually saw. He then began to experiment with these mental images of abstract geometrical designs and produced a body of work known for its "tessellations" (pronounced tess-el-AY-shuns). A tessellation refers to a checkered appearance, like a checkerboard. Looking at a typical Escher work of art that is tessellated, one sees some version of an intricately repeating geometric pattern. Throughout his life, Escher worked at producing images composed of geometric

Leone Alberti

Although trained in architecture, young M. C. Escher felt his artistic soul inspired and moved by the beauty of the Italian countryside. Nearly five hundred years earlier, another young man trained in architecture, Leone Battista Alberti (1404–1472), expressed his own artistry by using his mathematical skills. Alberti was born in Genoa, Italy, and received a mathematical education from his father. Although he later studied law, he would become a perfect example of what is commonly called a "Renaissance man," someone who does not specialize in any one field and has a broad range of interests and skills.

Besides being an architect and mathematician, Alberti was a musician, painter, sculptor, and writer of tragedies. He is best known, however, for writing the first book on the mathematical laws of perspective, titled *Della pittura e della statua* (On Painting and on Statuary). Although many artists studied and used mathematics during the early fifteenth century and did give some depth or realism to their paintings, it was not until Alberti showed his theoretical genius that the correct rules were stated that tell an artist exactly how to paint three-dimensionally.

Until Alberti, perspective, or the achievement of visual depth on a flat surface, was achieved mainly by a trial-and-error method. Artists knew that to the eye, objects in the distance appear smaller and that parallel lines seem to meet, but the question was always how much smaller and where should they meet to have a real, three-dimensional look? Alberti found the answer in geometry, and in his book he wrote that the first requirement for a painter should be that he know geometry. Using mechanical aids like pinhole cameras, grids, and string to locate the "vanishing point," he was able to establish the rules of perspective that allowed Renaissance painters to represent a three-dimensional world on a flat, two-dimensional surface. Since it was the goal of all Renaissance painters to faithfully depict the real world, they embraced Alberti's laws of perspective and, using geometry, were able to create a realistic-looking world.

elements, which divided a surface into regular patterns. He also found different ways of creating the illusion of three dimensions on a flat surface by exploring such individual techniques known as "infinity" and "impossible constructions."

M. C. Escher

After 1936, Escher began to use what could be described as his own version of a mathematical approach to produce his drawings. For example, in order to give the impression of something appearing infinite, he experimented first by trying to make his repeating figures appear vague or fainter as they approached the edges of the print. He then tried to make the figures slightly smaller as they approached the middle, or smaller as they moved toward the border. As he was always experimenting through his art with what is now recognized as a kind of mathematics, Escher produced highly individualized drawings that were part fantasy and part logic. His odd perspectives and sometimes "impossible worlds" caught viewers's eyes and brought them back for a second look. In many ways, his work often resembled puzzles, even if they were tessellations in the recognizable forms of humans or animals.

Although Escher was not a mathematician in any way, he became increasingly aware of the close connection between mathematics and his art. He began reading the mathematical work of his contemporaries, and eventually started corresponding with a few of them. During the 1950s and 1960s, as his fame continued to grow, he found that more and more of his admirers were mathematicians. This is not surprising since much of what Escher has done can be seen as concrete examples of certain mathematical concepts. Many teachers use Escher's art to show and explain the concepts of symmetry, infinity, and other more complex ideas. In the 1979 Pulitzer Prize–winning book, *Gödel, Escher, Bach: An Eternal Golden Braid*, Douglas R. Hofstadter marvels that Escher not only had the ability to visualize these strange worlds but to portray them as well. That is, he was able to actually draw them. Said Hofstadter, "Many have their origin in paradox, illusion, or double meaning," and behind each, "there is often an underlying idea, realized in artistic form."

A bridge between science and art

When World War II broke out in Europe in 1939, Escher moved back to the Netherlands where he would remain for the rest of his life. By the 1960s, the mathematical community had embraced Escher to the extent that his work was exhibited at some of their international conferences. During October 1960, Escher even lectured at the Massachusetts Institute of Technology in Cambridge, Massachusetts. As his health failed in the 1960s, however, he trav-

M. C. Escher

Up and Down, an example of
the work of M. C. Escher.
*Reproduced by permission of
Art Resource.*

M. C. Escher

elled less but continued to work. During July 1969, he made his last graphic work, a woodcut called *Snakes*. In 1972, he died. Unlike nearly any other artist, Escher's art was driven by an inborn or natural kind of understanding of complex mathematical concepts, and it was his rare genius to offer the world thousands of examples of art that form a bridge to connect the sometime distant worlds of science and art.

For More Information

Alberti, Leon Battista. *On Painting*. New Haven, CT: Yale University Press, 1956. Reprint, London: Susan Allix, 1999.

Cordon Art B. V. *The Official M. C. Escher Website*. http://www.mcescher.nl/ (accessed July 12, 2002).

Ernst, Bruno. *The Magic Mirror of M. C. Escher*. New York: Random House, 1976. Reprint, Amsterdam, Netherlands: Evergreen, 1994.

Escher, M. C. *The Magic of M.C. Escher*. New York: Harry N. Abrams, 2000.

Hofstadter, Douglas R. *Gödel, Escher, Bach: An Eternal Golden Braid*. New York: Basic Books, 1979.

IProject Online. *M. C. Escher Page*. http://www.iproject.com/escher/escherhome.html (accessed July 12, 2002).

Locher, J. L., ed. *M. C. Escher, His Life and Complete Graphic Work: With a Fully Illustrated Catalogue*. New York: H. N. Abrams, 1982.

"M. C. Escher." *Totally Tessellated*. http://www.abc.lv/thinkquest/tq-entries/16661/index2.html (accessed July 12, 2002).

O'Connor, J. J., and E. F. Robertson. "Leone Battista Alberti." *The MacTutor History of Mathematics Archive*. School of Mathematics and Statistics, University of St. Andrews, Scotland. http://www.groups.dcs.st-andrews.ac.uk/~history/Mathematicians/Alberti.html (accessed July 12, 2002).

O'Connor, J. J., and E. F. Robertson. "Maurits Cornelius Escher." *The MacTutor History of Mathematics Archive*. School of Mathematics and Statistics, University of St. Andrews, Scotland. http://www.groups.dcs.st-andrews.ac.uk/~history/Mathematicians/Escher.html (accessed July 12, 2002).

World of Escher. http://www.WorldOfEscher.com/ (accessed December 2001).

Ziring, Neal. *M. C. Escher Pages.* http://users.erols.com/ziring/escher.htm (accessed July 12, 2002).

M. C. Escher

Born 1933
Tupelo, Mississippi

African American algebraist and educator

Etta Zuber Falconer

Etta Zuber Falconer.
Reproduced by permission of Etta Zuber Falconer.

Etta Zuber Falconer is a distinguished mathematician who has devoted her entire career to increasing the number of African American students in mathematics and mathematics-related careers. As a teacher, scholar, and administrator at Spelman College in Atlanta, Georgia, she is recognized by her peers as being one of the most influential and respected leaders in mathematics and science education.

Meets mentors and role models

Etta Zuber Falconer was born in Tupelo, Mississippi, in 1933. She and her older sister Alice are the daughters of Walter A. Zuber, a physician, and Zadie L. Montgomery, a musician who attended Spelman College. She attended public schools in her hometown of Tupelo and graduated from George Washington High School in 1949. Always interested in science and mathematics, she enrolled at Fisk University, a historically black college in Nashville, Tennessee, to study chemistry, with mathematics as a minor. By her sophomore year, however, she decided that mathematics was much more interesting than chemistry, and she reversed her major and minor fields. At Fisk, she encountered two strong, successful teachers who served as her role models and mentors (trusted counselors) and who influenced her to set her sights high.

The first of the Fisk teachers, Lee Lorch (1915–), was chair of the mathematics department. Lorch encouraged Falconer and was instrumental in helping her focus on a career goal of becoming a mathematician. He advised Falconer to take graduate classes during her senior year and to attend graduate school, since as an eager student, Falconer had completed all the requirements for her major by her junior year. The second of the Fisk teachers, **Evelyn Boyd Granville** (1924–; see entry in volume 1), instructed Falconer for only one year, but was equally as inspiring.

Until late in her college career, Falconer had always assumed that she would graduate and use her degree in mathematics to teach high school. But her admiration for Lorch and Granville caused her to change that thinking. She realized she did not have to limit herself as a professional in her chosen field: Teaching mathematics at the college level was an achievable goal, even if being a woman—especially an African American woman—would instantly place her in the minority. In 1953, Falconer graduated *summa cum laude* (pronounced SOO-muh koom LAHW-dee; with highest honors) with a bachelor's degree in mathematics. She was also inducted into Phi Beta Kappa. Following Lorch's advice, Falconer entered the University of Wisconsin at Madison to pursue a master's degree in mathematics.

Out of her element for the first time

For Falconer, however, moving to Wisconsin would not be just like attending another school in another state. In 1954, American society was still racially segregated, meaning that African Americans were restricted to the use of their own separate institutions, like schools and churches, and facilities, like restaurants and restrooms. Although they were at times allowed to mix with whites, such instances were rare and were certainly the exception. As noted in *Black Women in Mathematics,* Falconer later remembered how strange a world she entered back then, saying, "Can you imagine what it was like for a nineteen year old black female from Tupelo, Mississippi, who had been immersed in segregation for all her life to attend the University of Wisconsin? I underwent a major culture shock...." For Falconer, graduate school meant her

Etta Zuber Falconer

first experience living and working in a nonsegregated environment. Although it was not segregated, there were no black professors nor any women professors on campus. Falconer was left to socialize with "students from Africa, a roommate from Thailand, and an office-mate from India...." For the most part, she found herself isolated academically and had little or no social interaction with her fellow students, most of whom were white males who did not take her seriously as a student.

One of her more unpleasant memories during her time there was when she taught a class for the mathematics department. This was required since she had received a teaching assistantship, which meant she had to teach in exchange for tuition and expenses. The first time she appeared before a group of white students to teach a college **algebra** (see entry in volume 1) class, she recalled them laughing as if her presence in front of the class were a joke. Despite this and other hostile situations, she managed to keep her respect and was able to show that she was indeed a qualified teacher. When she completed her work for the master's degree in mathematics in 1954, Falconer was given the opportunity to remain at Wisconsin and work toward her Ph.D. But Falconer had endured enough of the stress and hostility that went with her situation, and she decided to return to Mississippi.

Teaches, marries, and earns her Ph.D.

For the next nine years, she taught mathematics at Okolona Junior College in Okolona, Mississippi, thirty miles from her hometown of Tupelo. It was during this time that she met her husband, Dolan Falconer. They would have three children, Dolan Falconer Jr., an engineer, Alice Falconer Wilson, a physician, and Walter Falconer, also a physician. During these years, the National Science Foundation established a Teacher Training Institute to strengthen the nation's science curriculum by training college teachers—especially in mathematics—for four consecutive summers. Falconer was chosen to attend in the summer of 1962, and by the end of her third summer, she accepted an offer to direct the Institute and to attend the University of Illinois to work on her doctorate.

The entire Falconer family moved to Illinois in 1964, and she did very well at the Institute. However, when her husband was offered a position at Morris Brown College in Atlanta, Georgia, she decided to follow him south again. Shortly thereafter, she was hired to teach mathematics at Spelman College, a historically black liberal arts college for women in Atlanta (which her mother had attended). Falconer also decided to continue pursuing her own advanced degree, and while teaching at Spelman she enrolled at Emory University in Atlanta. In 1969, she was awarded her Ph.D. in mathematics from Emory.

A long career at Spelman College

Falconer has spent most of her professional life at Spelman College. She has held positions of instructor/associate professor (1965–71), professor of mathematics and chair of the mathematics department (1972–82), chair of the division of natural sciences (1982–90), director of science programs and policy (1990), and associate provost for science programs and policy (1991–present). She has also been the college's Fuller E. Calloway Professor of Mathematics since 1990.

This steady series of promotions and increased responsibilities have allowed Falconer to positively impact the lives of many young women in mathematics and the sciences. She has devoted her career to encouraging African American students, particularly women, to study mathematics and science. During her years at Spelman, she has become one of America's most productive and influential mathematics and science educators. She has generously shared her time, talent, and energy in pursuit of her lifelong commitment to helping young people learn and advance in the fields of mathematics and science.

Awards and activism

Falconer has received many awards in recognition of her work on behalf of the next generation of scientists and mathematicians. One of the most prestigious is the Louise Hay Award from the Association for Women in Mathematics (AWM) in 1995 for contributions to mathematics education. She also received the United Negro College Fund (UNCF) Distinguished Faculty Award (1986–87), Spelman's Presidential Award for Excellence in Teaching (1988), Spelman's Presidential Faculty Award for Distinguished Service (1994), the National Association of Mathematicians's (NAM) Distinguished Service Award (1994), the Giants in

Etta Zuber Falconer

Etta Zuber Falconer

Science Award from the Quality Education for Minorities (QEM) Network (1995), and an honorary doctor of science degree from the University of Wisconsin at Madison (1996). In addition, Falconer is one of the founders of the National Association of Mathematicians, which promotes the concerns of African American students and mathematicians, and the Atlanta Minority Women in Science network.

Throughout her life, Falconer has regularly demonstrated her commitment to improving mathematics education for her community and for all students. She has offered her students a nurturing environment and has contributed to their success by having high expectations for them while also building their self-confidence. Despite her busy schedule and the demands on her time, she continues to teach mathematics courses, knowing this will always keep her in touch with her students. She has honored her mentors by becoming an admired and respected mentor to many herself.

For More Information

Bailey, LaKiea. "Etta Falconer." In *Biographies of Women Mathematicians.* http://www.agnesscott.edu/lriddle/women/falconer.htm (accessed July 12, 2002).

Houston, Johnny L. "Etta Zuber Falconer." *MAA Online.* http://www.maa.org/summa/archive/falconer.htm (accessed July 12, 2002).

O'Connor, J. J., and E. F. Robertson. "Etta Zuber Falconer." *The MacTutor History of Mathematics Archive.* School of Mathematics and Statistics, University of St. Andrews, Scotland. http://www.groups.dcs.st-andrews.ac.uk/~history/Mathematicians/Falconer.html (accessed July 12, 2002).

Parker, Ulrica Wilson. "Etta Zuber Falconer." In *Notable Women in Mathematics.* Westport, CT: Greenwood Press, 1998, pp. 43–47.

Williams, Scott W. "Etta Zuber Falconer." *Black Women in Mathematics.* http://www.math.buffalo.edu/mad/PEEPS/falcon ner_ettaz.html (June 2001) (accessed July 12, 2002).

**Born October 29, 1656
Haggerston, England**

**Died January 14, 1742
Greenwich, England**

English astronomer and mathematician

Edmond Halley

E dmond Halley will forever be linked to the comet that bears his name, yet he was also a multitalented mathematician who made important contributions to several fields of science. Able to excel at both pure and applied mathematics, he was a pioneer of social **statistics** (see entry in volume 2) and is considered the founder of scientific geophysics. As one of the most respected scientists of his time, he was also a good friend of English mathematician and physicist **Isaac Newton** (1643–1727; see entry in volume 2), and helped publish Newton's greatest work.

Edmond Halley.
Courtesy of the Library of Congress.

Early success in astronomy

Edmond Halley (pronounced HAL-ee) was born in Haggerston, England, which is outside London and in the borough of Hackney. His father, Edmond Halley, was a rich merchant who sold soap and salt, and owned businesses in London as well as various other rental properties. His mother, Ann Robinson, died in 1672, the year before Halley entered college. Edmond had a younger sister, Katherine, and a younger brother, Humphrey. Halley's father spared no expense on his oldest son's education, and arranged for a tutor to teach Halley at home then sent him to St. Paul's School in

Edmond Halley

London. Halley entered Queen's College, Oxford, at the age of seventeen, where he first formally studied astronomy, a subject in which he had shown an early interest. His father encouraged this interest and purchased Halley several astronomical instruments, one of which was a twenty-four-foot-long telescope.

While a student at Queen's College, Halley began a regular correspondence with John Flamsteed (1646–1719), the Astronomer Royal at the Greenwich Observatory, and the older man guided Halley's progress in astronomy (although they would later have a falling out and Flamsteed would remain Halley's rival until his death). In 1676, Halley was able to join an expedition to Saint Helena to map the stars of the southern hemisphere. St. Helena is an island and a British colony located in the South Atlantic Ocean about one thousand miles off the west coast of Africa. When Halley returned in 1678, he published the first star charts of the southern hemisphere and dedicated it to King Charles II (1630–1685). Impressed with the published work of a twenty-two-year-old, the king issued an order that Oxford should grant Halley his master's degree. A year later, Halley was inducted into the prestigious Royal Society, to which all the leading scientists of England belonged. At the young age of twenty-three, Halley had already established what would be an ever-growing scientific reputation.

Halley's Comet

Halley's best-known scientific achievement was his theory on the movement of comets. Beginning in 1680 with a trip to Paris to work with Italian French astronomer Giovanni Domenico Cassini (1625–1712), Halley began his painstaking observational work studying a new comet that had appeared that year. This was not to be his hallmark comet, but his detailed observations led him to consider a new theory that comets traveled in elongated or stretched out elliptical (egg-shaped) orbits and made regular or periodic returns past Earth. In other words, that the sightings of comets in the past were of the same comets making their return.

One of the comets he observed and studied intensively was the comet of 1682. This was to become known as Halley's Comet when it reappeared in 1758 (sixteen years after his death) as he predicted. He also predicted that it would reappear roughly every seventy-five or seventy-six years thereafter (1835, 1910, 1986, and

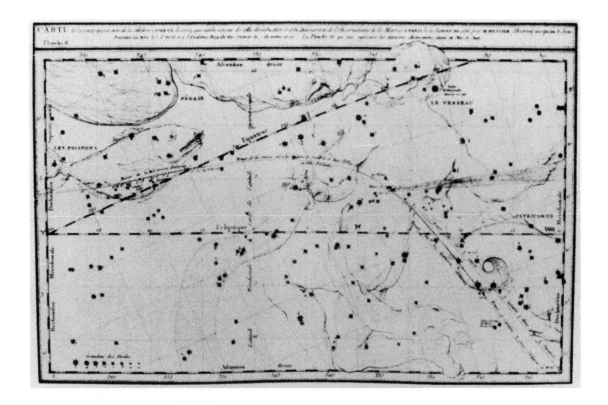

Celestial chart showing
constellations and the route of
Halley's Comet in January 1759.
Courtesy of the Library of Congress.

so on). To arrive at this prediction, Halley spent years searching
historical records and studying the movements of two dozen other
comets. After meeting and becoming friends with the great Isaac
Newton in 1684, Halley realized that comets were affected by the
gravitational influence of planets. It was Newton, of course, who
would make the scientific world aware of the universal principle of
gravitation with the 1687 publication of his great work,
Philosophiae Naturalis Principia Mathematica (The Mathematical
Principles of Natural Philosophy).

In 1705, after years of mathematical calculations and study, Halley
published a paper in which he first outlined his theory of elliptical
orbits for comets. He further stated the bold prediction that the
comet of 1682 would, as part of its natural, periodic cycle, follow a
path that would have it return past Earth for all to see during
December 1758. Although this paper and the book that followed
aroused some interest among astronomers and scientists, it was
not until the comet of 1682 did in fact reappear on December 25,
1758, that the entire scientific world took note. Halley's hope that
the world would remember the comet was first discovered by an

Halley's Comet, passing through
space in 1986.
*Reproduced by permission of the
Corbis Corporation.*

Englishman came true as the comet soon was named after him. More importantly, his successful prediction was viewed as proof or direct evidence of the correctness of Newton's theory of gravitation. With Halley's Comet, Newtonian mechanics became universally accepted.

Wide-ranging scientific contributions

Among Halley's other astronomical achievements were his work on the nature and number of stars in the universe, the distance of the Earth from the sun, and how to use the stars to navigate and find longitude at sea. In studying these astronomical questions, his mathematics was always essential. In fact, Halley showed great interest in early mathematics, and when he was appointed Savilian professor of **geometry** (see entry in volume 1) at Oxford in 1704, he began a translation of the seven books of the *Conics* by Greek mathematician **Apollonius of Perga** (c. 262–c. 190 B.C.E.; see entry in volume 4). This work took several years, and Halley even produced a reconstruction of the missing Book VIII.

In pure mathematics, Halley published seven papers ranging from higher geometry to **logarithms** (see entry in volume 2) and trigonometric functions. He also published papers in which he used mathematics to solve everyday problems, such as calculating a cannon's trajectory (the curved path it follows to its target) and computing the focal length of thick lenses. He also wrote papers on his archaeological (scientific study of the material remains of past human life and activities) research and on the nature of sunlight. He is considered by many to be the founder of scientific geophysics, which is the study of the physics of the Earth. He wrote on solar heating causing trade winds and monsoons, made the first meteorological (weather) chart containing real data, studied the tides, and contributed a theory on Earth's magnetism.

Halley was also a very practical man, and he carried out many experiments having to do with the sea. In addition to designing several different practical instruments, he designed a diving bell and a diver's helmet that were far ahead of anything of the time. He also established the mathematical law connecting barometric pressure with heights above sea level

Friedrich Wilhelm Bessel

One of the pioneers of nineteenth-century astronomy was a young man who left school at age fourteen and used his native mathematical talents to get a job in a merchant's counting house. Friedrich Wilhelm Bessel (1784–1846) was only able to learn Latin during his few years in school, but he discovered his own genius when he left school and took a job as an apprentice to a bookkeeper in Bremen, Germany. Since he was interested in the problem of navigating at sea, he taught himself navigation along with geography, mathematics, astronomy, and several foreign languages.

Bessel's talent was discovered by German astronomer Heinrich Wilhelm Matthaus Olbers (1758–1840), when the twenty-year-old Bessel sent Olbers his work on Halley's Comet. What Bessel had done was to recalculate the orbit of Halley's Comet using a set of two-hundred-year-old observations made by English mathematician and scientist **Thomas Harriot** (1560–1621; see entry in volume 3) in 1607. This work so impressed Olbers that he arranged for Bessel to eventually become a professional astronomer.

Bessel's talents were as much in mathematics as in astronomy, and he was able to combine the two to achieve major breakthroughs. In astronomy, Bessel is hailed as being the first to accurately determine the distance of a star from Earth, which is better known as the first stellar parallax. Bessel had to painstakingly measure for an entire year the parallax or shift in the position of a star that occurred as the Earth orbited the Sun. He also was the first to use the term "light years" as a vivid way of explaining great astronomical distances. In mathematics, Bessel is best known for his "Bessel functions," which he used to unlock the mystery of what are called "planetary perturbations" or a planet's slight drifting from its regular orbit. Using Bessel functions, scientists can accurately predict the existence of another planet or body, or determine the degree of planetary shift if the sizes of nearby bodies are known. Bessel functions have become a highly useful tool in physics and in engineering.

(showing to what degree pressure decreases as one climbs higher). Finally, Halley is recognized as one of the pioneers of keeping and using social statistics, as he used his mathematical skills in 1693 to show how mortality tables could be used as the basis for calculating life insurance payments. It was this practical sense that got him in trouble with church officials and other believers, for he took a mathematical and scientific approach to the question of the age of the Earth and decided that the Biblical estimate was far too low. He also got in trouble by trying to scientifically explain Biblical events, such as the flood in the story of Noah.

Long and illustrious career

Besides his position at Oxford as Savilian professor of geometry, Halley was made Astronomer Royal in 1720. This appointment demonstrated the high regard in which he was held. In 1682, Halley married Mary Tooke, the daughter of an auditor at the royal bank. They had three children, Katherine, Margaret, and Edmond. His daughters survived him, but his son, a naval surgeon, died a year before Halley. His wife had died five years earlier. Halley was described as a freethinker who had a lively sense of humor and who could be extremely diplomatic and charming when necessary. His colleagues considered him a friendly man who was always ready to offer support to younger scientists. His friendship with Newton is known to have played a major role in the publication of Newton's *Principia*, which many consider to be the greatest scientific work ever published. Were it not for Halley's close collaboration with Newton, as well as his steady encouragement and actual financial assistance, the book that "altered the whole course of physical science" might never have been written. It was Halley who used his own money to pay for the publication of Newton's great work.

For More Information

Calder, Nigel. *The Comet Is Coming!: The Feverish Legacy of Mr. Halley*. New York: Viking Press, 1980.

Girard, Linda Walvoord. *Earth, Sea, and Sky: The Work of Edmond Halley*. Niles, IL: A. Whitman, 1985.

Heckart, Barbara Hooper. *Edmond Halley, the Man and His Comet*. Chicago: Children's Press, 1984.

Hibben, Sally. "Edmond Halley." In *The Great Scientists*. Edited by Frank Magill. Danbury, CT: Grolier Educational Corporation, 1989.

O'Connor, J. J., and E. F. Robertson. "Edmond Halley." *The MacTutor History of Mathematics Archive*. School of Mathematics and Statistics, University of St. Andrews, Scotland. http://www.groups.dcs.st-andrews.ac.uk/~history/Mathematicians/Halley.html (accessed July 12, 2002).

O'Connor, J. J., and E. F. Robertson. "Friedrich Wilhelm Bessel." *The MacTutor History of Mathematics Archive*. School of Mathematics and Statistics, University of St. Andrews, Scotland. http://www.groups.dcs.st-andrews.ac.uk/~history/Mathematicians/Bessel.html (accessed July 12, 2002).

Ronan, Colin A. "Edmond Halley." In *Biographical Dictionary of Mathematicians*. New York: Charles Scribner's Sons, 1991, pp. 960–65.

Born January 8, 1942
Oxford, England

English theoretical physicist and writer

Stephen Hawking

Stephen Hawking.
Reproduced by permission of
AP/Wide World Photos.

Stephen Hawking has been called the most brilliant theoretical physicist since German American physicist **Albert Einstein** (1879–1955; see entry in volume 2). Hawking has spent his career studying such difficult and weighty subjects as the origin and fate of the universe, and has gained a worldwide following as an author and lecturer. As a pioneer in the study of black holes, Hawking regularly uses complex mathematics to provide proof for his theories.

An average childhood

Stephen William Hawking was born in Oxford, England, during World War II (1939–45). His parents were living in London at the time of his birth, but his mother was sent to the relative safety of Oxford, where she gave birth to Stephen. The family soon reunited in Highgate, north of London, where Hawking's father, Frank Hawking, worked. His father was a research physician who specialized in tropical diseases, and his mother, Isobel, was a college graduate and the daughter of a physician. After Stephen was born, she had two daughters, Mary and Philippa, and adopted a boy, Edward. The children were raised in a lively, talkative, and intellectual environment. Young Hawking began his education in Highgate, but when his father accepted a position at the Institute for Medical Research in Mill Hill, the family moved in 1950 to the cathedral town of St. Albans, twenty miles northwest of London.

Hawking then enrolled at St. Albans School and remained there through high school. As a young person, Hawking always knew that he lacked what he called "social graces." In the biography, *Stephen Hawking: A Life in Science* by Michael White and John Gribbin, young Hawking is described: "He was eccentric and awkward, skinny and puny. His school uniform always looked a mess and, according to friends, he jabbered rather than talked clearly, having inherited a slight lisp from his father."

For the longest time, Hawking had known that he wanted to be a scientist, and by his middle teens he had decided to study physics or mathematics. However, he was by no means a brilliant child, and his abilities in school did not have any real impact on teachers. It was at St. Albans, however, that he became inspired by a teacher to learn more mathematics, and as his interest grew, he began to make rapid strides. As he became more interested in studying subjects that were concerned with measurable quantities and objective reasoning, his mathematical skill increased and he was soon outdistancing everyone in class. He suddenly found himself able to get the highest grades without having to do hardly any homework. Hawking was unathletic and had a group of friends who shared his scientific interests. Together, they built a primitive computer in 1958 that actually worked.

Attends Oxford and Cambridge

Hawking's father never encouraged his son's interest in mathematics, mainly because he wanted him to attend his old school, University College, Oxford, and it did not offer a fellowship or grant in mathematics. In the spring of 1959, young Hawking took the scholarship examinations for Oxford with the aim of studying the natural sciences. Although he felt that he had not done well on this test, he was awarded a scholarship to Oxford where he would specialize in physics as part of his natural sciences education. At Oxford, Hawking's extraordinary abilities blossomed and soon became obvious to all. He seemed to be able to handle the most difficult problems without ever having to study. In 1962, Hawking received a first-class honors degree and set off for Cambridge

University to begin studying for an advanced degree in cosmology (the study of the nature of the universe).

By the time he was at Cambridge, Hawking was beginning to think seriously about some of the questions and themes that would concern him the rest of his life. One of the subjects that captured his attention, and which was not fully understood by science, was that of black holes. A black hole is a cosmic body that by its very nature can never be seen. One type of black hole was thought to be the remains of a collapsed star, which possessed such intense gravity that nothing could escape from it, not even light.

Develops Lou Gehrig's disease

At Cambridge, Hawking first noticed the early symptoms of a condition called amyotrophic lateral sclerosis (ALS), also known as Lou Gehrig's disease. This is a condition that affects the spinal cord and causes weakness throughout the body. The symptoms of this disease get progressively worse, and death often follows a few years after a diagnosis, when the chest muscles weaken to the point where the patient can no longer breathe. Hawking was diagnosed in early 1963, and his doctors predicted that he would not live long enough to finish his doctorate.

However, Hawking had two reasons to keep living. First, he was making excellent progress on his doctoral research. He later said, "although there was a cloud hanging over my future, I found to my surprise that I was enjoying life in the present more than I had before. I began to make progress with my research...." The other reason was that he had met a woman whom he wanted to marry. With the help of his doctors and his highly supportive parents—who designed their own therapy to help him—Hawking was able to continue to battle his disease. In 1965, with his condition under some control, Hawking married Jane Wilde, a language student at Cambridge. They would have two sons, Robert and Timothy, and a daughter, Lucy, before separating in 1990. In 1995, Hawking married his former nurse, Elaine Mason.

Formulates Hawking radiation theory

After earning his Ph.D. from Cambridge in 1965, Hawking obtained a fellowship in theoretical physics at Gonville and Caius College in Cambridge. There, he met and began a collaboration with English

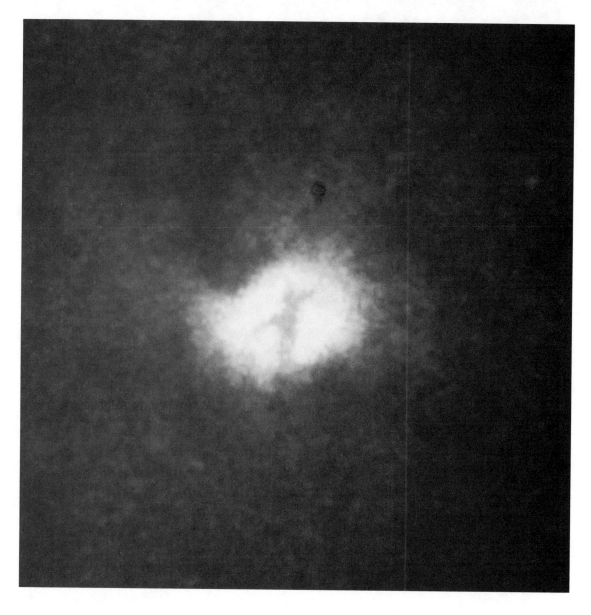

mathematical physicist Roger Penrose (1931–), who was also deeply interested in black holes and theories of space-time. Although only in his early twenties, Hawking had already acquired an impressive reputation, which, in 1968, led to an offer of a position at the Institute of Astronomy in Cambridge, which he accepted. By then, he and Penrose were using highly complex mathematics to apply the laws of thermodynamics (the physics that deals with the mechanical action or relations of heat) to black holes. This work resulted in their devising new mathematical techniques to study this area of cosmology.

The core of the Whirlpool galaxy M51. The ring of dust and gas is thought to surround and hide a giant black hole.
Courtesy of the National Aeronautics and Space Administration.

Around 1970, Hawking discovered what was a remarkable property of black holes. By using both Einstein's theory of general relativity (which states that gravity controls the universe and does so in a predictable manner) and quantum theory (which states that matter behaves randomly at the level of the atom and below), he was able to show that black holes can emit radiation. This was a challenge to the conventional notion that nothing can ever escape from a black hole. By 1973, he was able to prove mathematically that what he called mini-black holes actually give off particles and radiation (energy in the form of waves), after which they gradually evaporate and explode. This theory—that black holes are not in fact "black"—has since been accepted by most physicists, and "Hawking radiation" is the term used to describe these black hole emissions.

Seeks origin of the universe

In 1974, as evidence of his acceptance and success, Hawking became one of the youngest ever to be admitted to the Royal Society, a group in which all the leading scientists of England belonged. After spending a year as a Fairchild Distinguished Scholar at the California Institute of Technology in Pasadena, California, Hawking returned to England to continue his work toward a theory of the origin of the universe. This led him to question the big bang theory, which most scientists accept as explaining the origin of the universe. Hawking wondered whether there had ever been a real beginning to space-time (the big bang), or whether one universe simply gave birth to another without a beginning or an end. He also suggested that new universes might be born frequently through little-understood irregularities in space-time. At this level of cosmological physics, mathematics is essential, since the only proofs available are mathematical ones. Hawking found that his growing disabilities actually helped him visualize things more than others could, and he described this ability, saying he "tended to think in pictures."

Publishes highly popular book

In 1988, Hawking's *A Brief History of Time: From the Big Bang to Black Holes* was published. It immediately hit the best-seller lists in both England and the United States, staying there for several years and setting many records. Intended for a general audience, the book explains in simple language what centuries of people have thought about the nature of the universe and describes the evolu-

Roger Penrose, who worked with
Stephen Hawking in the
late 1960s.
*Photography by Anthony Howarth.
Reproduced by permission of Photo
Researchers, Inc.*

tion of his own thinking about the cosmos. The book's remarkable
success made Hawking an instant worldwide celebrity. In 1992,
the book was even made into a movie, directed by Errol Morris
(1948–). Subsequent books have also become best-sellers, further
demonstrating Hawking's unique ability to make his work accessible to those who have no background in science or mathematics.

Four years before *A Brief History of Time* was published, Hawking
suffered a major health crisis when he contracted pneumonia and

Stephen Hawking

Stephen Hawking (in wheelchair) and director Errol Morris attend the movie premiere of *A Brief History of Time* in 1992. *Reproduced by permission of the Corbis Corporation.*

had to be put on life support. A delicate throat operation saved his life but took away his ability to speak. Ever since, Hawking has spoken with a computerized electronic voice. Although he has long been confined to a wheelchair and can only move a few fingers of one hand, he has nurses, family members, and friends that attend to his needs. One of them stated that although his body has withered and almost wasted away, his mind is "capable of somersaulting through space and time." The awards, honors, and prizes he has received over the years is staggering. Besides receiving honorary degrees from several major institutions, he has received the Eddington Medal of the Royal Astronomical Society (1975), the Pius XI Gold Medal (1975), the Maxwell Medal of the Institute of Physics (1976), the Franklin Medal of the Franklin Institute (1981), the Gold Medal of the Royal Society (1985), the Paul Dirac Medal and Prize (1987), and the Britannica Award (1989).

Always a student of the history of science, Hawking seems to take special pride in his connection to two of the greatest mathematicians and physicists who ever lived. First, he often refers to the fact that his birth date coincides with the three-hundredth anniversary of the death of Italian astronomer, physicist, and mathematician **Galileo Galilei** (1564–1642; see entry in volume 1). Second, he takes great pride in his 1979 appointment as Lucasian Professor of Mathematics at Cambridge, since this position was held three centuries earlier by English mathematician and physicist **Isaac Newton** (1643–1727; see entry in volume 2). Hawking continues to study, research, and write, despite his condition.

For More Information

The Grolier Library of Science Biographies. Danbury, CT: Grolier Educational, 1997.

Hawking, Stephen W. *A Brief History of Time: From the Big Bang to Black Holes.* New York: Bantam Books, 1988.

O'Connor, J. J., and E. F. Robertson. "Stephen William Hawking." *The MacTutor History of Mathematics Archive.* School of Mathematics and Statistics, University of St. Andrews, Scotland. http://www.groups.dcs.st-andrews.ac.uk/~history/Mathematicians/Hawking.html (accessed July 12, 2002).

Professor Stephen Hawking's Homepage. http://www.hawking.org.uk/ (accessed July 12, 2002).

White, Michael, and John Gribbin. *Stephen Hawking: A Life in Science.* New York: Plume/Penguin Books, 1993.

Stephen Hawking

Born June 14, 1935
Metz, France

Died October 28, 1989
Oak Park, Illinois

American logician and educator

Louise Hay

Louise Hay earned a worldwide reputation for her research in mathematical **logic** (see entry in volume 2). She was a founding member of the Association of Women in Mathematics, and in 1980, was the only female head of a major, research-oriented university mathematics department in the United States. She is regarded as an ideal educator and a person whose life story serves as an inspiration to every student.

Early life fleeing the Germans

Louise Schmir Hay was born in Metz, France, which is directly east of Paris and near the borders of Luxembourg and Germany. Her father, Samuel Szmir (pronounced SCHMUR), had emigrated from Poland to France and was in the clothing business. Her mother, Marjem Szafran, was also from Poland, and she died when her daughter was only three years old. Her father married Eva Sieradska later that year, and altogether, the family consisted of three children: Gaston, born in 1933, Louise, born in 1935, and Maurice, born in 1943. Since the Szmir family was Jewish, they spent much of World War II (1939–45) trying to evade capture by the Germans who had a national policy of exterminating anyone of Jewish origin. In March 1944, the family decided to split up and young Louise and her older brother managed to get to Switzerland (a neutral country during the war), where they remained for the next year. When Germany was driven out of France by the Allied forces (the people who fought against Germany in the war) in early 1945, she and her brother returned to France. The entire family then emigrated to the United States in 1946 and changed their last name to the more pronounceable Schmir. Hay's father owned a delicatessen in New York from 1950 to 1965.

Chooses mathematics rather late

Hay attended William Howard Taft High School in the Bronx, New York, and until then she had no particular interest in any sort of mathematics. She actually did much better in her nonmathematical subjects. However, in the tenth grade, she was lucky enough to have a geometry teacher who turned her completely around on mathematics. This was because he believed in teaching the logic of the subject rather than simply teaching his students by the theorem-proof method. In other words, he expected his students to understand what they were doing when they wrote up a proof. Young Louise found this logical aspect of mathematics much more appealing to her than its regular numerical aspects, and she became very interested in what he was teaching.

Hay's teacher suggested that she read a book about non-Euclidean **geometry** (see entry in volume 1), and he also arranged for her to have a tutor in mathematics. She found the book fascinating, and she used the topic as the basis for a project she did in her senior year that eventually won her third prize in the Westinghouse Science Talent Search. This prize enabled her to win a scholarship to attend Swarthmore College in Swarthmore, Pennsylvania, where she would major in mathematics. The Westinghouse award also led to summer jobs at the National Bureau of Standards and a part-time job at the Moore School of Electrical Engineering at the University of Pennsylvania, which helped support her through college.

Mathematical logic

At Swarthmore, Louise majored in mathematics and always intended to go to graduate school to study mathematical logic. However, in the 1950s, there were not a great many mathematics

Louise Hay

departments that offered an advanced degree in mathematical logic. As a branch of mathematics, mathematical logic is almost closer to philosophy in that it focuses on one characteristic that all mathematical branches have in common—deductiveness (pronounced dee-DUCK-tiv-ness). This means that all mathematical propositions are demonstrated from axioms or postulates (certain things that are accepted or given). What each of these has in common, no matter how different they are, is that conclusions are demonstrated, step-by-step, from these things that are given. In this way, the basic characteristic of mathematics consists of demonstrating a conclusion from a hypothesis. From this point of view, therefore, the purest mathematics would not pay attention to *what* is being demonstrated but rather focus only on the *process* of demonstration or deduction. This deductive logic aspect of mathematics is what Louise liked best about mathematics.

Encounters roadblocks to advanced degrees

At the end of her junior year at Swarthmore, Louise married John Hay, whose field was experimental psychology. When she graduated and received her bachelor's degree in 1956, she joined her husband who was already at Cornell University in Ithaca, New York, where they both received teaching assistantships. After two years there, Louise Hay left Cornell to go to Oberlin College in Oberlin, Ohio, where her husband had accepted a teaching job. When she left Cornell in 1959, she received her master's degree but had no concrete plans to get her Ph.D. After a year at Oberlin, she worked at the Cornell Aeronautical Laboratory for another year, only to move to Massachusetts where her husband had obtained a job at Smith College.

In Massachusetts, Hay taught part-time at a junior college for one year, after which she taught for three years at Mount Holyoke College in South Hadley, Massachusetts. At this point, the successful launch and orbiting of the Soviet satellite *Sputnik,* and the emergence of a challenging Soviet space program suddenly made Americans aware that there was a shortage of professional people in mathematics and the sciences. By this time, Hay was more than ready to begin work on her Ph.D. in mathematical logic, and there were greater opportunities than ever before. However, it was about this time that she became pregnant with her first child, Bruce, and she naturally assumed that her dream of getting a Ph.D. in the subject she loved was about to be put on hold again.

Changes her life

However, during the spring of 1963, several things happened to change her life. First, she read *The Feminine Mystique,* by women's rights activist Betty Friedan (1921–), and began to think long and hard about her decisions to always give priority to her husband's career at the sacrifice of her own. At the same time, she also met algebraist Hannah Neumann (1914–1971), who told Hay of her own professional difficulties. Neumann had interrupted her studies to have two children, was evacuated from England as an "enemy alien," despite the fact that her husband was in the British Army, eventually returned to school, had three more children, and still received her Ph.D. Inspired by Neumann, Hay decided to stay on at Cornell (where her husband had again gone for the summer), while her husband returned to Smith College in the fall.

Despite attacks of anxiety and insomnia, Hay landed a job as a research assistant, took three courses, passed her preliminary examinations, and began work on her Ph.D. thesis. Ready to begin writing her thesis, she found herself pregnant with twins, Philip and Gordon. Although this delayed her writing and presenting her thesis because they were born prematurely, by the end of 1965 she had received her Ph.D. from Cornell. After staying home for a year, Hay was rehired by Mount Holyoke as an assistant professor, and was able to spend one year on a National Science Foundation fellowship at the Massachusetts Institute of Technology (MIT). During this time, she remembers hiring countless babysitters, and endured the disapproval of many who thought mothers should never work. When her marriage broke up and she and her husband divorced in 1968, she found that having a Ph.D. made a considerable difference in her ability to advance herself. That same year, she accepted a position at the University of Illinois at Chicago as associate professor. In 1970, she married Richard Larson, a colleague in the mathematics department.

Becomes department head

In the late 1960s and early 1970s, Hay published a series of papers on mathematical logic that were considered highly original. As noted in *Biographies of Women Mathematicians Web Site,* she remembered "the peculiar thrill of briefly knowing a sliver of mathematical truth that *nobody* else knows." In 1975, Hay was promoted to pro-

Louise Hay

Louise Hay

fessor at Illinois. The year before, she became involved with the beginnings of a new professional organization, the Association for Women in Mathematics, and was among the first to be invited to address that group. In 1979, she became acting head of the mathematics department at Illinois, and headed that department from 1980 until 1988—the only woman in the country at the time to head a university department. As an administrator, she was very popular, democratized the department, made strong appointments, and helped change the name to include mathematics, statistics, and computer science. In 1974, Hay was diagnosed with cancer, which she battled until it recurred in 1988. Hay died a year later.

Shortly after Hay's death, the Association for Women in Mathematics created an award in her honor, which is given to a woman who has made an outstanding contribution to mathematics education. Hay regularly supported and encouraged women students at her school, and always believed that her experiences could be instructive to others. As noted in *Biographies of Women Mathematicians Web Site,* a year before she died, she said that if there is a moral tale to how she became a mathematician, it is, "that the sources of inspiration and opportunities to change your life can come unexpectedly and should not be ignored; and that you should not neglect the dictates of your own career, taking some risks if necessary, since you never know what the future will bring."

For More Information

Green, Judy, and Jeanne Laduke. "Women in American Mathematics: A Century of Contributions." In *A Century of Mathematics in America, Part II.* Edited by Peter Duncan. Providence, RI: American Mathematical Society, 1989, pp. 379–98.

Hay. Louise. "How I Became A Mathematician." *Biographies of Women Mathematicians Web Site.* Agnes Scott College. http://www.agnesscott.edu/lriddle/women/hay.htm (accessed July 12, 2002).

O'Connor, J. J., and E. F. Robertson. "Louise Schmir Hay." *The MacTutor History of Mathematics Archive.* School of Mathematics and Statistics, University of St. Andrews, Scotland. http://www.groups.dcs.st-andrews.ac.uk/~history/Mathematicians/Hay.html (accessed July 12, 2002).

Born c. 65
Alexandria, Egypt

Died c. 125
Alexandria, Egypt

Greek geometer and engineer

Hero of Alexandria

Hero of Alexandria is the most important ancient authority on mechanical devices. His **geometry** (see entry in volume 1) deals mostly with problems of computing length, **area** (see entry in volume 1), and **volume** (see entry in volume 2), and his most noted mathematical contribution was his method for approximating **square roots** (see entry in volume 2). His work contributed to a scientific foundation for engineering and land surveying.

Hero of Alexandria.
Reproduced by permission of the Corbis Corporation.

Little known of his life

Until the 1930s, there was much dispute concerning what year Hero of Alexandria was born. In fact, no one was even certain of the century in which he lived. Also called Heron and nicknamed "the mechanic" or "the machine man," Hero of Alexandria remained a mystery as a person. However, in 1938, the question about approximately when he lived was settled by a researcher who noted that an eclipse of the moon described by Hero as taking place at a very specific time and place corresponded to one that occurred in 62 C.E. This astronomical fact at least placed him in the correct century. There are other indications that he lived during that time period, as Hero himself quotes Greek geometer

Hero of Alexandria

Archimedes of Syracuse (287–212 B.C.E.; see entry in volume 1), who lived long before him. Further, Hero is not mentioned in any other written source earlier than 300 C.E., in a book by Greek mathematician Pappus of Alexandria (c. 290–350).

Most scholars agree that Hero probably lived and worked in Alexandria, most likely at its famous university or museum, which was still a center for research and learning. By Hero's time, Egypt's days of glory were long gone, and it had become simply another province of the Roman Empire. Despite its decline, Alexandria was still a very large and sophisticated city that drew on the intellectual traditions of Egypt, Babylonia, and Greece. It is partly because of this that scholars are not sure whether Hero was Egyptian or Greek.

Continues a long tradition

Despite the lack of historical records on Hero's personal life, a relatively large amount of his scholarly writings have survived. One thing his writing shows is that he was well-educated, probably not self-taught, and that he was very familiar with the writings of his predecessors (the people who came before him). If he, indeed, was educated in Alexandria, where he ended up teaching, then he certainly had access to the Great Library's collection of several hundred thousand books and texts. Also, since he was a specialist in both science and technology, it can be assumed that he studied all branches of learning in Alexandria, especially mathematics, physics, and mechanics (the science that deals with energy and forces and their effects on bodies).

An unusually large number of Hero's books—thirteen—have endured through the centuries. They can be divided into two classes: the mathematical, or geometric, and the mechanical. Scholars believe all his books were written as texts for students or as manuals for technicians. None of the books are particularly original, nor was this Hero's intent as he deliberately drew on the work of his predecessors. In fact, this is probably the major distinguishing characteristic of his writing—that he gathered much of his knowledge from previous writers whose work he knew very well. In doing this, Hero did scholarship a great favor since quite often the work of these older writers did not survive and the work is only known because of Hero.

Mathematical work

In his six mathematical works, Hero always stressed practical application over pure theory. To him, mathematics was a valuable tool that one needed to achieve anything practical. His most famous mathematical work, *Metrica,* is a three-volume book on how to calculate and divide area and volume, and in it he defines geometry as "the science of measuring land." In this work, "Hero's formula" is found. This famous formula is used to calculate the area of a **triangle** (see entry in volume 2) from the length of its three sides. Among its many uses, the formula provided surveyors of his day with a method of determining the area of land lots. Although named after him, like most of Hero's work, scholars believe it comes from a much earlier source. Hero knew all the mathematical knowledge of his time, so it is no surprise that he was the first Greek to develop a method of finding a good approximation of a number's square root.

Three of his other mathematical works serve as an introduction to geometry, a catalog and explanation of 133 geometrical terms, and a consideration of solid geometry. Solid geometry includes two types of geometry. One part is traditional, Euclidean geometry, which involves the study of three-dimensional figures. The other part, plane geometry, deals with two-dimensional figures. Where plane geometry works with lines, angles, **circles** (see entry in volume 1), and triangles, solid geometry deals with shapes that have length, width, and depth—in other words, things that are real and part of one's everyday world. Solid geometry is therefore used by every engineer, designer, and builder, among many others.

The other two publications of Hero's are works of practical measurement, as when he provides a method for calculating the seating capacity of a theater or for determining the number of jars a ship could store. Hero is also thought to be the first Greek mathematician to use systematic geometrical terminology (the names or technical terms for things) and symbols. Many historians think that Hero was not a scholar who summarized the work of others but that, in fact, he was as intelligent as **Euclid of Alexandria** (c. 325–c. 270 B.C.E.), the Greek geometer and logician who is considered the founder of geometry.

Mechanical work

As good a mathematician as he was, Hero may have been an even better "mechanic." His contributions to the field of mechanics are

Hero of Alexandria

Hero of Alexandria

wide-ranging, and he is thought to have achieved considerable fame in his day for some of his inventions. For the most part, these inventions focused on the practical, such as his device to keep **time** (see entry in volume 2) with a water clock and his contraption that used compressed air to catapult (hurl) things against an enemy. Perhaps his most famous mechanical design was for the "aeolipile" (pronounced ee-OH-luh-pile), which used steam to rotate a hollow sphere to which two bent tubes were attached. When water was boiled in the sphere, the steam escaped from the tubes, which started the sphere whirling about. This has been described by many as the world's first steam engine.

As with his mathematical books, Hero passed down a great deal of the technology that he already knew. Since he did such a good job summarizing the work of his predecessors, scholars consider him the ultimate authority on ancient Greek and Roman technology. In his seven mechanical books, with appropriate names like *Pneumatica* and *Automata,* he sometimes has serious and theoretical sections in which he discusses subjects like the existence of vacuums in nature and the five simple machines. Simple machines refers to the basic tools, known since ancient Greece, that are used to make work easier. There are five: the lever, the wheel and axle, the pulley, the inclined plane, and the screw. Some include the wedge as a sixth, but most consider it to be a variation of the inclined plane.

Overall, however, Hero's mechanical books offer a wide range of highly technical information of a very practical nature. For example, besides discussing machines powered by compressed air and steam pressure, Hero addressed such topics as mirrors and surfaces that reflect light; instruments for surveyors; the design and use of catapults for warfare; the use of toothed wheels for heavy lifting; and even designs for making miniature mechanized puppet theaters. One of his more curious devices is what might be considered the world's first slot machine. Used in Egyptian temples, this machine dispensed something like holy water when a person deposited a coin. Hero also designed a mechanical bird that could sing, temple doors that could open automatically, and a water fountain powered by compressed air. Apparently, Hero and his contemporaries saw only the playful aspects of steam and compressed air and did not try to use them for any real practical purposes.

Hero's reputation in mathematics and mechanics grew as technology advanced, especially in the last two centuries. He is now consid-

Two ancient military men pull the levers of a catapult. Among Hero of Alexandria's efforts in mechanics was his work on catapults for warfare.
Reproduced by permission of the Corbis Corporation.

ered to be much more than someone who just described ancient devices and told how they should be built. His efforts preserved a great deal of ancient information that had come down to him from earlier writers whose works had been lost. Even though he did not build all of the models he described, they presented engineers and technicians with interesting design ideas and problems, which only grew in importance as the world became more industrialized. Today, Hero's reputation is high, and scholars consider him a rare ancient who displayed an almost modern sense and feeling for mechanics, as well as a deep understanding for the practical uses of mathematics.

For More Information

Drachmann, A. G. "Hero of Alexandria." In *Dictionary of Scientific Biography.* Edited by Charles Coulston Gillispie. New York: Charles Scribner's Sons, 1973, pp. 306–9.

Franceschetti, Donald R., ed. *Biographical Encyclopedia of Mathematicians.* New York: Marshall Cavendish, 1999.

"Hero of Alexandria." *Historical Graphics and The History of Hydraulics Book Collection.* IIHR—Hydroscience & Engi-

Hero of Alexandria

neering, College of Engineering, University of Iowa. http://www.iihr.uiowa.edu/products/history/hoh/hero.html (accessed July 15, 2002).

Hughes, J. Donald. "Hero of Alexandria." In *The Great Scientists*. Edited by Frank N. Magill. Danbury, CT: Grolier Educational Corporation, 1989, pp. 49–53.

O'Connor, J. J., and E. F. Robertson. "Heron of Alexandria." *The MacTutor History of Mathematics Archive*. School of Mathematics and Statistics, University of St. Andrews, Scotland. http://www.groups.dcs.st-andrews.ac.uk/~history/Mathematicians/Heron.html (accessed July 15, 2002).

Born 1675
Anglesey (now Gwynedd), Wales

Died July 3, 1749
London, England

Welsh-born English geometer and educator

William Jones

The life and career of William Jones are a good example of how a person can be recognized as a contributor to the history of mathematics without being a great mathematician or making any sort of original contribution. Jones is remembered today as being the first person to use the symbol π to represent the **ratio** (see entry in volume 2) of the circumference of a **circle** (see entry in volume 1) to its diameter, known as **pi** (see entry in volume 2). He is also well-known because of his close association with English mathematician and physicist **Isaac Newton** (1642–1727; see entry in volume 2).

Early years unremarkable

William Jones was born in the part of Anglesey (pronounced ANG-gull-see), Wales, known as Llanfihangel Tw'r Beird. Wales is one of four countries making up the United Kingdom, and it is located on the southwest part of England. Although his father's name was John George, William became known as William Jones when he followed the old Welsh custom of translating his father's first name into a last name, making it into Jones (from John). His mother's name was Elizabeth Rowland. As a youngster in school, Jones showed enough ability and promise to attract the attention

An Inexact Slice of Pi

When William Jones selected the Greek letter "p" (π) as the symbol for pi (because it stands for and is the first letter of the word "perimeter"), he was probably as intrigued and as fascinated by the concept of pi as mathematicians are today, or as the Egyptians and Chinese were thousands and thousands of years ago. Pi is the ratio of the circumference of a circle to its diameter. Amazingly, that ratio is always the same, no matter the size of the circle. This fact alone must have been startling when first discovered. While the Egyptians knew that this ratio was a little more than 3, today the calculation is known to be a little more than 3.14. However, that is not its exact value since pi is an infinite **decimal** (see entry in volume 1). That means that, in decimal form, pi can never finish since it simply goes on and on, with no end. Simply, despite today's advanced computing power, no repeating pattern for pi has ever been discovered. In addition, pi cannot be written as a **fraction** (the ratio of two integers; see entry in volume 1). Pi is therefore an irrational number since that is how an irrational number is defined.

The concept of pi has been surrounded by many interesting facts and strange stories. For instance, in 1897, the state legislature of Indiana tried to simplify things by making it a state law that pi would have the value of 3. Although the House passed the bill, the Senate did not vote on it, and it was effectively killed. Today in the United States, National Pi Day is celebrated by enthusiasts, teachers, and students on March 14 of every year. This date, 3/14, corresponds to the simple value of pi (3.14). Those who want to be even more precise, celebrate it at exactly 1:59 on 3/14 (since pi to five places is 3.14159). Although not every mathematics class celebrates—or even knows about—National Pi Day, the Exploratorium in San Francisco, California, can be counted on to celebrate National Pi Day every March 14.

of a local landowner named Bulksley of Baron Hill, who then became his patron. A patron is someone who uses his money or influence to help another individual; in the seventeenth century, the patronage of a wealthy person was necessary to a farmer's son who wanted to succeed. The helpful landowner eventually found him a position in the countinghouse (accounting firm) of a London merchant.

Travel and teaching

The job at the countinghouse took place on a ship that travelled to the West Indies, a group of islands in the Caribbean between North and South America. Jones served his company at sea and also taught mathematics to the men on board ship from 1695 to 1702. While at sea, Jones realized that he had a talent for teaching. After his voyages were over, he returned to London and began a new career as a mathematics tutor.

Jones must have been an excellent teacher, since he was hired by some of England's powerful families. One of his young pupils, Philip Yorke (1690–1764), went on to become lord chancellor, a very powerful cabinet position in the English government. The chancellor liked and respected Jones, and eventually invited him to work and travel with him. Jones also was connected to the Royal Society of London (the first scientific society), having tutored its president, George Parker, as well as Parker's father, Thomas. For many years, Jones lived at Shirburn Castle with the Parker family, and it was there he met and married Maria Nix, the daughter of a London cabinetmaker. They would have two sons and a daughter.

Writes mathematical works

In 1702, Jones wrote his first book, *A New Compendium of the Whole Art of Navigation*. Having just returned from the sea, he wrote it as an instructional manual for sailors. It applied mathematics to astronomy (which was important for calculating one's position at sea and for navigating by the stars). Following that, he wrote *Synopsis Palmariorum Matheseos: or, A New Introduction to the Mathematics*. Published in 1706 in London, this book caught the attention of Newton and his good friend, English astronomer **Edmond Halley** (1656–1742; see entry in volume 4). Although it was intended to be a book for beginners in mathematics, it discussed Newton's own invention, calculus, and included an overview of recent advances in mathematics.

Introduces the symbol for pi

The concept of pi, that there is a constant ratio between the circumference of a circle and its diameter, was known in a practical way as long as four thousand years ago. However, it was Greek mathematician **Archimedes of Syracuse** (287 B.C.E.–212 B.C.E.; see entry in volume 1), who after thoroughly analyzing and calculating the value of this ratio, stated that it was a little over three. Actually, he determined that a circle's circumference is always 3.14 times its diameter, no matter the size of the circle. Knowledge of this important mathematical rule or formula allows certain tasks to be performed quickly and efficiently.

When Jones discussed the concept of pi in his 1706 book, he used the symbol π to represent this important ratio. It is thought that he chose π because it is the first letter of the Greek word for perimeter, which is *perimetros*. Perimeter is also another word for the circumference of a circle. Pi is the sixteenth letter of the Greek alphabet, and it is often represented by the Greek symbol π for that letter. The mathematical use of the symbol π for pi did not become standard practice until Swiss mathematician **Leonhard Euler** (1707–1783; see entry in volume 1) adopted it in 1737. Following his use, it was accepted and used by all mathematicians.

William Jones

William Jones corresponded
frequently with
fellow mathematician Isaac
Newton (right).
*Reproduced by permission of the
Corbis Corporation.*

Close touch with Isaac Newton

After the publication of his second book, Jones was often in close
touch with Newton, and the two corresponded a great deal. Newton apparently thought so highly of Jones that he gave him access
to his manuscripts and allowed Jones to publish a few of his works
on calculus. In 1712, Jones was elected a fellow of the Royal Society of London and eventually served as its vice president. During
his lifetime, Jones managed to collect the private papers or manuscripts of several mathematicians, including Newton, but after his

death in 1749, the papers became so mixed together that it took years to separate out what had been done by Newton and what was the work of others. In fact, this task was not completed until sometime in the 1970s. Overall, the farmer's son who tutored the children of the rich and corresponded with the great Isaac Newton lived to see his name forever linked to a universal symbol.

For More Information

Baron, M. E. "William Jones." In *Biographical Dictionary of Mathematicians.* Vol. 2. New York: Charles Scribner's Sons, 1991, pp. 1182–83.

Burton, David M. *The History of Mathematics: An Introduction.* Boston: McGraw-Hill, 1999, p. 192.

Eves, Howard. *An Introduction to the History of Mathematics.* New York: CBS College Publishing, 1983, p. 87.

Hoffman, Joseph E. *The History of Mathematics to 1800.* Vol. 2. Totowa, NJ: Littlefield, Adams and Co., 1967.

O'Connor, J. J., and E. F. Robertson. "William Jones." *The MacTutor History of Mathematics Archive.* School of Mathematics and Statistics, University of St. Andrews, Scotland. http://www.groups.dcs.st-andrews.ac.uk/~history/Mathematicians/Jones.html (accessed July 15, 2002).

Born 1048
Nishapur, Persia (now Iran)

Died 1131
Nishapur, Persia

Persian mathematician, astronomer, and poet

Omar Khayyám

Omar Khayyám.
Reproduced by permission of the
Corbis Corporation.

Omar Khayyám was one of the most brilliant figures of Islamic civilization. A mathematician, astronomer, philosopher, and poet, he is best known in the West for his collection of poems, the *Ruba'iyat*. He was famous during his lifetime for his reform of the Islamic calendar, and his chief contribution to mathematics was his solution of cubic equations through **geometry** (see entry in volume 1).

Finds a patron

Known in the West as Omar Khayyám (pronounced KY-om), his full name was Gheyás od-Dín Abū ol-Fatḥ 'Omar ebn Ebrahīm ol-Khayyāmī. Born in Nishapur, Khorāsān, Persia (now Iran), his given name was Umar, while "ol-Khayyāmī" meant "tentmaker." This may have been his father's or his family's trade. Few facts are known of Khayyám's early years, although most agree that he was educated in the sciences and philosophy in his hometown of Nishapur. However, some think he may have also been schooled in Balkh, which is now in Afghanistan. Wherever he was educated, it is likely that he became a tutor, although teaching was a full-time job that would have kept him away from his scholarship, so it is also likely that he actively sought a patron. In eleventh century Persia, scholars were either wealthy or they were attached in some way to a patron, a person usually in the royal court, who would support them and their research. The main problem with such a system was that a scholar's fate was usually linked to that of his patron, so if the political or military situation shifted and his patron fell out of power, the scholar usually suffered in some way as well.

In 1070, Khayyám traveled to Samarkand, one of the oldest cities in Central Asia (now in Uzbekistan), where he obtained the help and support of Abu Tahir, a prominent judge. While there, Khayyám used his time well and produced several works, one of which was a book on **algebra** (see entry in volume 1). Translated, its title is *Treatise on Demonstration of Problems of Algebra,* and it would become his most famous mathematical work. As young Khayyám's reputation for scholarship grew, he was soon invited by the sultan and ruler of the city of Isfahan, Malik-Shāh (1055–1092), to come to that city's royal observatory and supervise a team of astronomers he had assembled.

Leads calendar reform

Khayyám accepted the sultan's invitation, and for the next eighteen years, he oversaw the construction of an observatory and led some of the best astronomers of his time. Under his guidance, they compiled astronomical tables and conducted important work on the reform of the solar calendar then in use in Persia. A solar calendar is one that is based on the solar or seasonal year, the time it takes the Earth to go once around the Sun. Having an accurate solar calendar was important to rulers who used a calendar to make revenue collections, such as taxes, and for other administrative matters that had to take place at certain times.

In 1079, Khayyám managed to quantify the length of the solar year. The figure he arrived at, 365.24219858156 days, is amazingly accurate given that this figure was obtained nearly one thousand years ago. Using these calculations, Khayyám then devised a solar calendar that included 8 separate leap years (of 366 days, or one extra day) in every cycle of 33 years. Although his system was slightly more accurate than the Gregorian calendar, it proved a bit more difficult to use. The Gregorian calendar was introduced in

Omar Khayyám

the West by Pope Gregory XIII (1502–1585) in 1582 to correct errors in the long-used Julian calendar, and over time, it became adopted by most countries. It is still in use worldwide.

Omar Khayyám

A new patron

During his lifetime, Khayyám lived in what is best described as an unstable military empire. In 1092, political events brought an end to his peaceful and productive years at Isfahan. In that year, sultan Malik-Shāh, who had originally invited Khayyám to his observatory, was murdered by a group of terrorists known as the Assassins. Khayyám had not only lost a patron, but he acquired new and powerful enemies as well. Further, the observatory lost its funding and the calendar reform was halted. Khayyám himself came under attack from traditional Muslims who felt that he was too open and too questioning, especially in religious matters.

Despite the unsteady atmosphere, Khayyám remained in Isfahan and tried to convince the sultan's successors to provide support for science and education. Finally, in 1118 he was invited by Malik-Shāh's third son, Sanjan, to travel to Merv (now the city of Mary in Turkmen), which he had made the capital of his empire. Beginning with Khayyám's arrival, that city became a great center of Islamic learning, and Khayyám was again able to do productive scientific and even literary work. He would remain in that city until near the end of his life when he returned to his native Nishapur.

Mathematical contributions

In Khayyám's book on algebra, he offered what is now regarded as his great contribution to mathematics—his method of solving cubic equations. He was the first to offer a general theory of cubic equations. Until Khayyám, the one thing that algebra could not do was solve equations of the third degree or those in which cubes, like x^3, appeared. These equations were much more complex than simple linear (first degree) or quadratic (second degree) equations. First degree equations (x^1) are called linear equations because any graph made of them always results in a straight line. Equations become more complicated when the unknown factor x is raised to a power of 2 (x^2, or x times itself). These second degree or quadratic equations show their complexity by graphing as a curve. They naturally become even more complicated when the unknown x is raised to a power of 3.

Regiomantanus

Regiomantanus.
*Reproduced by permission of the
Corbis Corporation.*

As Omar Khayyám helped reform the Islamic calendar for the Sultan, four hundred years later, another well-known mathematician and astronomer was summoned by his leader to help reform the old Julian calendar of the West. Since the Christian observance of Easter was based on the flawed Julian calendar introduced by Roman emperor Julius Caesar (100 B.C.E.–44 B.C.E.), by the time of the Renaissance that holy day was gradually drifting away from its original spring observance (which was tied to the Jewish Passover). Because of this, Pope Sixtus IV (1414–1484) summoned the German mathematician/astronomer who called himself "Regiomontanus," meaning King's mountain.

Regionmontanus's real name was Johann Müller (1436–1476), and he had established an excellent reputation in both of his fields. His 1474 book, *Calendarium,* brought him to the pope's attention, and in 1475 the pontiff summoned Regiomontanus to Rome to begin work on a new calendar. However, Regiomontanus died in Rome in 1476 before he could accomplish very much. There are conflicting accounts as to how he died—some say he died of the plague and others argue that he was poisoned. Real calendar change in the West was not achieved until over a century later (1582) under Pope Gregory XIII. He reigned from 1572 to 1585, and it is after him that the Gregorian calendar is named.

In working toward a solution, Khayyám continued the tradition of Greek algebra, which used geometry to map algebraic problems. To do this, however, he had to imagine cubes and even complicated parallelepipeds (pronounced pear-uh-lell-uh-PYE-pehds; prisms whose bases are parallelograms) instead of just the squares and rectangles of first or second degree equations. Although he relied on geometry to solve equations, he did not use a straightedge and compass, but arrived at his solution by using conic sections (called this because they are the result of a plane intersecting a cone). Thus, he solved cubic equations by intersecting a parabola (pronounced puh-RAH-bo-luh; bowl-shaped curve) with a **circle** (see entry in volume 1). This was a remarkable achievement. Khayyám also studied the parallel postulate (something that is given or true) of Greek geometer **Euclid of Alexandria** (c. 325 B.C.E.–c. 270 B.C.E.), and laid the groundwork for an entirely new way of doing geometry.

Omar Khayyám

Poet and philosopher

Khayyám is known to have mastered history, philosophy, law, and even medicine, in addition to astronomy and mathematics. He also wrote on the theory of music and even authored some religious works (possibly to clear himself of the charge of atheism, the belief in no god). Today, his popularity in the West is based on his work as a poet, specifically as the author of the *Ruba'iyat*. This collection of four-line poems has become famous for its romantic, almost musical verse. It was first translated into English in the mid-1800s and became an instant success in America and England. In this work, Khayyám's poetry deals with the great questions of life—death, eternity, reality, time, and the purpose of human existence. Many people find his poems truly profound. Twentieth century mathematical logician and philosopher Bertrand Russell (1872–1970) said Omar Khayyám was the only man he knew of who was both a poet and a mathematician.

For More Information

O'Connor, J. J., and E. F. Robertson. "Omar Khayyam." *The MacTutor History of Mathematics Archive*. School of Mathematics and Statistics, University of St. Andrews, Scotland. http://www.groups.dcs.st-andrews.ac.uk/~history/Mathematicians/Khayyam.html (accessed July 15, 2002).

Reimer, Luetta. *Mathematicians Are People, Too: Stories from the Lives of Great Mathematicians*. Palo Alto, CA: Dale Seymour Publications, 1994.

Youschkevitch, A. P., and B. A. Rosenfeld. "Al-Khayyami." *Dictionary of Scientific Biography*. Edited by Charles Coulston Gillispie. New York: Charles Scribner's Sons, 1973, pp. 323–31.

Zahoor, A. "Omar al-Khayyam." http://users.erols.com/zenithco/khayyam.html (accessed July 15, 2002).

Born February 1698
Kilmodan, Argyllshire, Scotland

Died June 14, 1746
Edinburgh, Scotland

Scottish geometer and physicist

Colin Maclaurin

C olin Maclaurin is considered the outstanding British mathematician of the eighteenth century. Besides being a brilliant
mathematician who solved many problems in **geometry** (see entry
in volume 1) and applied physics, he is best known for being the
first to publish a systematic or logical proof of the theorems of
English mathematician and physicist **Isaac Newton** (1643–1727;
see entry in volume 2).

Colin Maclaurin.
*Reproduced by permission of the
Corbis Corporation.*

Mathematical prodigy

Colin Maclaurin (pronounced muh-KLOHR-un) was born in
Kilmodan, Argyllshire, Scotland, the son of a well-known minister, John Maclaurin, who was also a scholar and a very learned
man. The youngest of three sons, his brothers were John, who
would follow in his father's footsteps and become a minister, and
Daniel, who would die at a young age. Unfortunately, Maclaurin's
father died six weeks after his son's birth, and his mother died
when he was nine years old. His mother had moved the family to
Dumbarton once her husband died so that her boys would receive
a good education. After his mother died, Maclaurin went to live
with his uncle, Daniel Maclaurin, who raised and cared for him.

Colin Maclaurin

His uncle Daniel was a minister and raised his nephew to appreciate nature and believe in God.

Young Maclaurin showed such ability at a young age that he was able to enter the University of Glasgow at the age of eleven. Although he planned to study religion and become a minister, he was greatly influenced by Scottish mathematician Robert Simson (1687–1768), who was a professor at Glasgow, and within a year, he abandoned his plans and took up mathematics. Even before he met Simson, however, Maclaurin had become interested in mathematics on his own, having picked up a copy of *Elements,* by Greek geometer **Euclid of Alexandria** (c. 325 B.C.E.–c. 270 B.C.E.) and quickly mastered the first six of its thirteen books. Simson taught him the geometry of ancient Greece as well as the natural philosophy or science of Isaac Newton.

It is not clear exactly when Maclaurin received his degree from Glasgow University. But what is truly amazing is the subject of his thesis. Maclaurin was awarded a master of arts degree when he publicly and successfully defended his thesis, "On the Power of Gravity." Written by a teenager, this paper was a sophisticated defense of Newton's work, and contained advanced mathematical ideas that could be understood by only the best mathematicians. By the time he was nineteen years old, Maclaurin had accepted a position at Marischal College in Aberdeen as professor of mathematics. Although he was extremely young to be given such a position, ten days of examinations and competition had demonstrated that he was the best candidate.

Grand tour of Europe

After receiving his professorship at Marischal College, Maclaurin twice visited London where he made contact with the scientific circles in that capital and eventually was able to meet with Newton and Martin Folkes (1690–1754), who would much later become president of the Royal Society of London, one of the oldest and most prestigious scientific groups. At this time, Maclaurin was already at work on what would become his first book, *Geometria organica.* This was published in 1720 with the approval of Newton, who had reviewed it. This work would prove many of the theorems (statements or formulas obtained from a more general principle) that Newton had proposed, as well as solve several other important prob-

Colin Maclaurin

In his college thesis, teenage student Colin Maclaurin wrote of the work of English mathematician and physicist Isaac Newton (left). *Reproduced by permission of Archive Photos, Inc.*

lems in geometry. *Geometria organica* became a major contribution since, for many of his theorems, Newton had offered no proofs.

In 1722, Maclaurin agreed to leave Scotland and accompany the oldest son of one of the king's more important diplomats on a tour of Europe. He would also function as the boy's tutor. A tour of Europe in order to broaden and complete an education had become a tradition among the upper class. It appears that Maclaurin, however, either did not inform his college of his plans or did not obtain its permission, but simply left with the boy and began traveling. The two went first to Paris, where Maclaurin made it a point to meet with French mathematicians. Later, while living in Lorraine, France, Maclaurin wrote a paper, "On the Percussion of Bodies," that won him a prize from the French Academy of Sciences in 1724. Although the two did travel to other countries, they spent most of their time in France. In fall 1724, however, his young pupil suddenly died, and Maclaurin returned to Scotland, having been gone almost three years.

Moves to Edinburgh

Since he no longer could reclaim his old position at Marischal College because of the unusual circumstances under which he left, Maclaurin sought and eventually was given the chair of professor of mathematics at the University of Edinburgh in 1725. Most agree that Maclaurin owed his appointment to the direct intervention of Newton, who not only wrote a letter to the school in which he strongly recommended the young man, but he also offered to contribute to Maclaurin's salary if the university could not pay him fully. Maclaurin would spend the rest of his productive career at Edinburgh.

With his new position, Maclaurin soon moved into Scotland's inner circles and was made a fellow of the Royal Society and the Philosophical Society. In 1733, he married Anne Stewart, the daughter of the solicitor general in Scotland, and they would have seven children. Only five of these children, two boys and three girls, would reach adulthood.

Newton's calculus

While Maclaurin wrote on several mathematically related topics such as the eclipse of the Sun, the theory of tides, and the statistics involved in the insurance business, his major contribution was his 1742 book called *Treatise of Fluxions*. Written as a reply to a published criticism of Newton's invention of the fluxions or calculus, this work became the first logical and systematic explanation of Newton's methods. In this two-volume work, Maclaurin explained Newton's calculus in great detail and laid out an excellent geometrical framework to support it. His book was so persuasive that it actually had a negative effect on the future of calculus in Britain

The Calculus Dispute

Gottfried Leibniz.
Courtesy of the Library of Congress.

As a teenager, Colin Maclaurin had written a sophisticated defense of the mathematical work of English mathematician and physicist Isaac Newton (1642–1727). He later became the primary explainer and defender of Newton's invention of calculus. History would eventually show that calculus was created independently by Isaac Newton and German logician and philosopher Gottfried Leibniz in the late seventeenth century, but there was considerable disagreement over whose invention came first.

Calculus is the means of calculating the way quantities vary with each other, and not just the quantities themselves. In 1665, young Isaac Newton created what he called the "method of fluxions" or the calculus. He did not make his discovery known, however, until he published an account in 1693. Leibniz came up with his independent creation around 1673, but he too waited and published only in 1684 (but before Newton). Both men had supporters who made a career of disputing the claims of the other, and eventually the dispute became an international one.

In fact, however, the two great men and their inventions were actually very different. Leibniz was a worldly man of great and very different talents. He was a major philosopher as well as a practicing diplomat, mathematician, and physicist. Newton was a solitary figure and somewhat of a social misfit who, despite being one of the great geniuses, spent considerable time and brainpower as master of the mint (in charge of coinage) and searching for chemical formulas to turn objects into gold. Where Leibniz would die neglected, poor, and even dishonored in his own country, Newton was a legend in his time and was given a state funeral. The irony is that although Newton was indeed first with the calculus, his method was so clumsy and difficult that all mathematicians eventually adopted the Leibniz method.

While the dispute over whose invention was first may have been pointless, calculus would go on to have immense implications for science, especially physics, as it made possible for the first time the analysis of movement. With it, physicists were finally able to represent motion geometrically using curves.

since it influenced British mathematicians to neglect any other versions that were not strictly Newton's.

During this time, German logician and philosopher **Gottfried Leibniz** (1646–1716; see entry in volume 2) had invented his own version of calculus, which would prove to be much more convenient than Newton's. However, British mathematicians would not use Leibniz's or any other developments in calculus that were made in the rest of Europe for the next three generations due to Maclaurin's influential work.

Defended against rebellion

As a person, Maclaurin was always known as an excellent teacher who genuinely cared for his students. He was also known for his outstanding kindness. As a true and loyal Scotsman, Maclaurin took an active role in defending his city of Edinburgh against the Jacobite army in 1745. The Jacobite movement was a British rebellion against the Crown. In September 1745, Jacobite rebels attacked Edinburgh, and Maclaurin worked endlessly to organize the city's defenses and to maintain its walls against the enemy. He worked tirelessly, planning and supervising its defense, and eventually drove himself into a state of exhaustion. Nonetheless, the city finally fell and he was forced to flee to York, England.

When it became clear that the Jacobites were not going to occupy Edinburgh, Maclaurin returned to Edinburgh in November 1745. On the cold trip back, Maclaurin fell from his horse and he was never able to recover his health. He died in Edinburgh in 1746 at the age of forty-eight.

For More Information

"Colin Maclaurin." *Gazetter for Scotland.* http://www.geo.ed.ac.uk/scotgaz/people/famousfirst829.html (accessed July 15, 2002).

Franceschetti, Donald R., ed. *Biographical Encyclopedia of Mathematicians.* New York: Marshall Cavendish, 1999.

O'Connor, J. J., and E. F. Robertson. "Colin Maclaurin." *The MacTutor History of Mathematics Archive.* School of Mathematics and Statistics, University of St. Andrews, Scotland. http://www.groups.dcs.st-andrews.ac.uk/~history/Mathematicians/Maclaurin.html (accessed July 15, 2002).

Scott, J. F. "Colin Maclaurin." In *Biographical Dictionary of Mathematicians.* New York: Charles Scribner's Sons, 1991, pp. 1637–40.

Born c. 1619
Schleswig-Holstein, Denmark (now Germany)

Died January 14, 1687
Paris, France

Danish mathematical astronomer

Nicolaus Mercator

Although not well-known today, Nicolaus Mercator was a man of many talents. The author of textbooks on astronomy and trigonometry, he also explained how **logarithms** (see entry in volume 2) can be used, and gave a mathematical introduction to astronomy. He wrote on calendar reform, explained the physical geography of Earth, and made an important distinction between **rational and irrational numbers** (see entry in volume 2).

Trained in Denmark

Nicolaus Mercator (pronounced murr-KAY-tur) was born as Nikolaus Kauffman in Eutin, Schleswig-Holstein, Denmark, which is now part of western Germany. His father, Martin Kauffman, was a schoolmaster at Oldenburg in Holstein, and it is believed that his son was educated at his father's school. His father died in 1638, but his son was able to attend the University of Rostock (in northeast Germany), and he received a degree from that institution in 1641. After spending a short time studying in Leiden, Holland, he returned in 1642 to Denmark and accepted a position as a faculty member at Rostock. During his time there, he wrote several textbooks on astronomy and trigonometry.

Trigonometry is the study of the relationships between the sides and the angles of **triangles** (see entry in volume 2) and the calculations based on them.

Nicolaus Mercator

In 1648, Mercator moved to the University of Copenhagen, where he taught and wrote more texts on astronomy and trigonometry. After teaching there for six years, he left when the university was forced to close because of an outbreak of the plague (a disease that is highly contagious and deadly). Before he left Denmark for good in 1660, he wrote textbooks that explained the application of logarithms to trigonometry and offered a plan to reform the calendar. Logarithms are numbers known as **exponents** (see entry in volume 1) that are used to express repeated **multiplications** (see entry in volume 2) of a single number. It is believed that Mercator's work on the calendar attracted the attention of Oliver Cromwell (1599–1658). Cromwell was an English soldier and statesman who successfully led the parliamentary forces against the king, and who eventually became lord protector of England, Scotland, and Ireland from 1653 to 1658.

Establishes himself in England

It is unclear whether Mercator received an invitation from Cromwell to come to England, but Mercator moved there in 1660 and immediately began having financial problems. He was unable to find a teaching position with a university, and took up private tutoring to earn some money. He would never stop earning a living this way. He also changed his name and became known as Nicolaus Mercator. Doing what many non-English scholars had done before him, he "Latinized" his German last name. Because of this, he is sometimes confused with Flemish geographer Gerardus Mercator (1512–1594), who, a century before, invented the technique of using curved lines on maps to designate longitude (the vertical lines on a map) and latitude (the horizontal lines on a map). Interestingly, he, too, changed his name, as his real name was Gerhard Kremer.

Although Mercator was never affiliated with any university in England, he was able to make the acquaintance of several of England's more prominent mathematicians. Fortunately, he had come to England already known as the inventor of a new marine chronometer (pronounced kruh-NAHM-eh-tur; a clock that works on a ship at sea and can be used to determine longitude). On the basis of this achievement and his newfound friends, Mercator was able to be elected a member of the newly formed and prominent Royal Society of London. The person most instrumental in nominating him was English mathematician and naturalist Sir Robert Moray (c. 1608–1673), who was one of the Society's founders and who acted as Mercator's patron (supporter).

Mathematical contributions

Before he ever came to England, Mercator had produced an impressive body of work. His 1651 book, *Trigonometria sphaericorum logarithmica,* explained how to use logarithms in trigonometry. Two other works published the same year, *Cosmographia* and *Astronomia* explained the physical geography of the Earth and gave a mathematical introduction to astronomy. In both of these texts, Mercator made the distinction between rational and irrational numbers. A rational number is a ratio (meaning division) that is symbolically represented as a **fraction** (see entry in volume 1) or as one number over another, as long as the bottom number is not a zero. An irrational number cannot be expressed as a ratio of two numbers, and when carried out, it simply goes on forever. In 1653, Mercator published his *Rationes mathematicae* in which he likened rational and irrational numbers to the concepts of harmony and dissonance (pronounced DIS-uh-nince; discord or clashing notes) in the study of music. That same year, he wrote a paper in which he suggested that the calendar should be changed. He offered a plan that would have 365 days made up of 12 unequal months. Although he received a good deal of criticism for these ideas, today a similar system of months with unequal number of days is used.

In 1664, he published his first book in England, a work titled *Hypothesis astronomia nova,* which discusses the laws of planetary motion put forth a generation earlier by German astronomer **Johannes Kepler** (1571–1630; see entry in volume 2) in light of his own theories. Mercator was known to have corresponded with English mathematician and physicist **Isaac Newton** (1642–1727; see entry in volume 2), and some historians believe that Mercator's theories may have been used by Newton in the formulation of his own laws of motion. Despite all this work in astronomy, Mercator was best known as a mathematician whose 1668 work on logarithms, titled *Logarithmotechnia,* made him an authority.

Nicolaus Mercator

Although Mercator had a sharp mathematical mind and was a skillful inventor, little is known of his personal life. He was described by his friends as a soft-tempered man, short in stature, with black hair, and dark eyes. He was also known to be shy and not outgoing, and this may account for him being lesser known than some of his more outgoing mathematical colleagues. In addition to his mechanical skills, he also possessed considerable engineering talent. In 1683, he accepted a commission from Jean-Baptiste Colbert (1619–1683), the minister of finance to King Louis XIV (1638–1715) of France, to plan the waterworks and fountains in Versailles, France (which would become the king's palace in northern France as well as the seat of French government for more than one hundred years). After moving to Versailles, Mercator eventually had a falling out with his financial supporters and died in Paris at the age of sixty-eight.

For More Information

O'Connor, J. J., and E. F. Robertson. "Nicolaus Mercator." *The MacTutor History of Mathematics Archive.* School of Mathematics and Statistics, University of St. Andrews, Scotland. http://www.groups.dcs.st-andrews.ac.uk/~history/Mathematicians/Mercator_Nicolaus.html (accessed July 15, 2002).

Westfall, Richard S. "Nicolaus Mercator (Niklaus Kauffman)." *Galileo Project: Catalog of the Scientific Community.* http://es.rice.edu/ES/humsoc/Galileo/Catalog/Files/mercator_nic.html (accessed July 15, 2002).

Whiteside, D. T. "Nicolaus Mercator (Kauffman, Niklaus)." In *Biographical Dictionary of Mathematicians.* New York: Charles Scribner's Sons, 1991, pp. 1699–1701.

Born June 13, 1928
Bluefield, West Virginia

American algebraist and game theorist

John Nash

John Nash is a highly original mathematician whose early, pioneering work in the field of game theory won him a Nobel Prize in 1994. His genius was stilled for nearly thirty years however, as he suffered from a debilitating mental disease. This disease gradually went away in the mid-1980s and he was able to resume his mathematical work once again.

John Nash.
Reproduced by permission of the Corbis Corporation.

Early signs of genius

John Forbes Nash Jr. was born in Bluefield, West Virginia, in the southernmost part of the state at the Virginia border. He was named after his father, an electrical engineer who joined the Appalachian Power Company after serving in France during World War I (1914–18). His mother, Martha Virginia Martin, was known as Virginia and had a university degree. She majored in languages and had been a Latin teacher for ten years before marrying and having children. Two years after her son John was born, she had a daughter named Martha. Although Nash and his sister were raised during the Great Depression (1929–41)—a prolonged economic slump that put many people out of work—his father kept his job and the family did not suffer in any way. In fact, they lived in a very comfortable house down the street from a country club.

John Nash

Almost from the beginning, Nash was different from most children. He preferred reading books to playing with other children, he was very good at chess, and he could do things like whistle an entire piece of classical music. In elementary school, some of Nash's teachers thought he was a slow child, and his mathematics teacher even said he was having difficulty understanding. This was surprising, since his parents took great interest in their children's education and regularly worked at home with both of them. In fact, young Nash was probably bored in school, and already looking for new and different ways to solve problems and learn. In addition, he lacked some social skills, likely leading his teachers to think of him as slow. However, while he may have been bored in school and had few friends, by the time he was twelve, he was conducting science experiments in his room.

At fourteen, Nash became seriously interested in mathematics when he read *Men of Mathematics,* by Eric T. Bell (1883–1960). Just a few years before, a young woman in California had read the same book and also became inspired to make mathematics her career. This was American mathematician **Julia Bowman Robinson** (1919–1985; see entry in volume 3), who would become the first female mathematician elected to the National Academy of Sciences. Whatever worked for her also worked for Nash, and this book of biographies of successful mathematicians stimulated Nash to study very difficult mathematics, at which he found he was very successful. Nash finally experienced the exciting and challenging side of mathematics that he had never enjoyed at school.

Discovers his genius

In 1945, Nash entered the Carnegie Institute of Technology (now Carnegie-Mellon University) in Pittsburgh, Pennsylvania. He chose chemistry as a major but was persuaded to switch to mathematics during his freshman year after a mathematics professor discovered his remarkable talent. He was, in fact, so good in school that he received his bachelor's degree in only two years and immediately began his graduate studies. During this time, Nash remained a mostly solitary person with few friends. He did not help his social situation when he acted odd and sometimes had childish tantrums.

However, by 1948, Nash had demonstrated such great mathematical potential that he was offered fellowships from several top graduate schools, including Harvard and Princeton.

In 1948, Nash decided to enter the doctoral program at Princeton, whose Institute for Advanced Study was the home of such renowned mathematicians as German American physicist and mathematician **Albert Einstein** (1879–1955; see entry in volume 1) and Hungarian American mathematician John von Neumann (1903–1957; see entry in volume 2). Einstein's theory of relativity had transformed twentieth century physics, and von Neumann's mathematical work helped to develop the modern **computer** (see entry in volume 1) and the hydrogen nuclear bomb.

Entering Princeton in September 1948, Nash immediately impressed his professors, although he did not attend many of their lectures. In fact, much of his graduate education was an exploration of his own mathematical genius, as he would develop a topic on his own and then try to solve a particular problem in an entirely unique way. Although he was proving himself intellectually at twenty among much more mature people, he was still a loner who often acted oddly. Some of his classmates thought he was not only unusual but obnoxious as well.

Mathematical principles of game theory
One of the more astounding and remarkable things about Nash is the fact that his slender twenty-seven-page doctoral thesis written at Princeton while in his early twenties eventually helped establish an entirely new field of mathematics called game theory. Later, the thesis would also serve as the basis for his creativity being recognized internationally with a Nobel Prize in Economic Science. Game theory was a brand new field, having been invented only four years before by von Neumann and his colleague, German American economist Oskar Morgenstern (1902–1977). It was von Neumann's belief that the type of mathematics normally developed for the physical sciences was not useful for fields like economics. He argued that such subjects in the social sciences necessarily involved human actions, which, in turn, were based on both choice and chance. He therefore proposed that a different mathematical approach was needed that could analyze strategies and take into account the independent choices of two or more participants.

John Nash

John Nash

In 1944, von Neumann and Morgenstern founded a new branch of mathematics they called game theory.

Game theory became a popular concept after World War II (1939–45) when the United States and the Soviet Union were engaged in a nuclear arms race. Game theory was applied to military and diplomatic strategy and was used to predict the possible outcome of future conflicts. Policy makers found that von Neumann's and Morgenstern's work offered them some logical and mathematical rules about rivalries. When Nash considered this new field, he realized that von Neumann's theory was only useful for pure rivalries or what were called "zero-sum" games. That is, his theory applied only to a situation in which two participants were involved in a winner-takes-all situation. Nash knew that real life was much more complicated, so he chose a topic for his doctoral thesis that von Neumann had not solved. Nash focused on "game" situations in which both parties had a mutual gain. In his brief but brilliant thesis, he essentially laid out what could be called the new mathematics of competition. This immediately turned game theory into a powerful new tool for analysts of all types, especially economists. Since economics always involves some form of competition, Nash's ideas were used to analyze everything from business competition to trade negotiations.

Suffers mental breakdown

Nash received his doctorate on his twenty-second birthday, and after teaching a short time at Princeton, he took a job with the RAND Corporation where his founding thesis, entitled *Non-Cooperative Games,* naturally made him the leading expert in the field he had helped establish. He soon returned to school in 1952 when he accepted a teaching position at the Massachusetts Institute of Technology (MIT). It was there he met Alicia Larde, an El Salvadoran physics student, whom he married in February 1957. By the autumn of 1958, the couple was expecting their son, but Nash's mental state had become very disturbed. Although Nash's few years at MIT had been very productive mathematically, he had been slowly going downhill mentally, and had more and more problems that were accompanied by strange behavior. Only after he was hospitalized in a psychiatric institution was Nash's odd behavior and personality problems recognized as the symptoms of a medical disorder. Nash was diagnosed as a paranoid schizo-

John Nash

John Nash and his ex-wife
Alicia arrive at the Academy
Awards show on March 24, 2002.
A Beautiful Mind, a movie
about Nash, won the Oscar for
best picture.
*Reproduced by permission of the
Corbis Corporation.*

phrenic (pronounced PEHR-uh-noyd skitz-oh-FREH-nik). In
the 1950s, this disease was considered treatable by psychotherapy
(a therapy in which a person's unconscious motives are examined).
In fact, some of his psychiatrists thought his mental state was
somehow related to his wife's pregnancy.

Recovers after lost years

For the next twenty or more years, Nash was basically lost to his
work and to his family. Although his wife eventually divorced him,

John Nash

she never remarried and allowed him to live with her and their son, supporting them both as best she could. For years, Nash was allowed to wander around the Princeton campus where nearly all knew and respected him, despite the fact that he had become a very sad individual. In the late 1990s, researchers learned that schizophrenia is a brain disease, although its causes are not known. Its victims suffer in many ways, as they often struggle with hearing voices. They are often fearful, and have trouble telling the difference between what is real and what is a delusion or fantasy. Nash often felt that he was being spied on or hunted, and he regularly searched numbers for what he thought contained hidden messages. In such a mental state, he could neither have a normal life nor think logically enough to do any mathematical work.

Despite what appeared to be a hopeless situation, Nash slowly began to actually recover after over twenty years. His former wife believed that what he needed was a quiet home life in the community he knew well, as well as the love and support of family and friends. She firmly believes that it was this, rather than any drugs or treatment, that accounts for what many believed was a miraculous remission. Amazingly, Nash eventually was able to work on serious mathematics again, and this was a true test of his recovery. As if his victory over schizophrenia was not enough, Nash received the 1994 Nobel Prize in Economic Science, which he shared with two others. Thus, at the age of sixty-six, the man who was at one time considered to be, as noted in *The MacTutor History of Mathematics Archive*, "the most promising young mathematician in the world," received international recognition for the genius he displayed as a twenty-two-year-old in a twenty-seven page dissertation. Nash's story is so dramatic and uplifting, that Sylvia Nasar wrote a biography on his life called *A Beautiful Mind*, which was made into a successful film in 2001 that won an Academy Award for best picture.

For More Information

Casti, John L. *Five Golden Rules, Great Theories of 20th-Century Mathematicians—and Why They Matter*. New York: John Wiley & Sons, Inc., 1996.

"John Nash: Genius, Nobel and Schizophrenia." *Popular-Science. net.* http://www.popular-science.net/nobel/nash.html (accessed July 23, 2002).

"John F. Nash, Jr.—Autobiography." *Nobel E-Museum*. http://www.nobel.se/economics/laureates/1994/nash-autobio.html (accessed July 23, 2002).

Kuhn, Harold W., ed. *The Essential John Nash*. Princeton, NJ: Princeton University Press, 2002.

Nasar, Sylvia. *A Beautiful Mind: A Biography of John Forbes Nash, Jr., Winner of the Nobel Prize in Economics, 1994*. New York, NY: Simon & Schuster, 1998.

Nasar, Sylvia. "The Lost Years of a Nobel Laureate." *The New York Times* (November 13, 1994).

O'Connor, J. J., and E. F. Robertson. "John Forbes Nash." *The MacTutor History of Mathematics Archive*. School of Mathematics and Statistics, University of St. Andrews, Scotland. http://www.groups.dcs.st-andrews.ac.uk/~history/Mathematicians/Nash.html (accessed July 15, 2002).

John Nash

**Born March 5, 1574
Eton, Buckinghamshire, England**

**Died June 30, 1660
Albury, Surrey, England**

English clergyman and arithmetician

William Oughtred

*William Oughtred.
Reproduced by permission of
Hulton/Archive by Getty Images.*

William Oughtred was a mathematical pioneer who experimented with many different algebraic symbols. Two of his notations—the *"x"* for **multiplication** (see entry in volume 2) and the symbol "::" for proportion—continue to be used today. He also invented the earliest form of the slide rule.

Pursued mathematics on his own

William Oughtred (pronounced AW-tred) was born in Eton, Buckinghamshire, England, in 1574. His father, Benjamin Oughtred, was a scrivener (someone who teaches writing) at Eton School and taught William, his oldest son, arithmetic. Young Oughtred was a good enough student at Eton to enter King's College in Cambridge when he was only fifteen years old. He received his bachelor's degree in 1596 and his master's degree in 1600. Although very little mathematics was taught at either Eton or King's College, Oughtred developed a passionate interest in that subject and studied it on his own at night after finishing his other school work. This impressed English writer and biographer John Aubrey (1626–1697), who knew him well. As noted in *The Mac-Tutor History of Mathematics Archive,* Aubrey wrote that Oughtred "slept but little. Sometimes he went not to bed in two or three nights, and would not come down to meals till he had found out the question." He also described Oughtred as being a little man with black hair and black eyes, and being "full of spirit."

Becomes a minister and a teacher

In 1603, Oughtred was ordained as an Anglican priest (the same as an Episcopal minister), and the following year he began running a parish in Shalford. In 1610, he took a position with higher pay and responsibility at a parish in Albury, where he would stay until he died. Despite all the duties he had running a parish, Oughtred continued his after-hours habit of working on mathematics. He also began to privately tutor promising young mathematics students. Since he felt he was paid enough as a priest, he did not charge his students and even let them live in his house for free. Among his many pupils, the most famous were English mathematician John Wallis (1616–1703), English architect and astronomer Christopher Wren (1632–1723), and English astronomer Seth Ward (1617–1689).

Keys to arithmetic

While tutoring Lord William Howard, the young son of Thomas Howard (c. 1585–1646), the earl of Arundel, Oughtred began to write a book on **algebra** (see entry in volume 1) and arithmetic. The earl of Arundel became a supporter of his and encouraged him to publish his work once it was finished. Although this book was not even one hundred pages long when published in 1631, it contained almost all that was known about algebra and arithmetic to that point. Known by its shortened Latin title, *Clavis mathematicae* (The Keys to Mathematics), the book included a description of Hindu-Arabic notation, as well as **decimal** (see entry in volume 1) **fractions** (see entry in volume 1) and algebra. Oughtred may have only intended his book to be a reference text for his students, but it quickly received favorable attention from mathematicians in England and later from those throughout Europe. Because of this, Aubrey wrote that Oughtred was more famous abroad than he was at home, and that several major figures went to England just to meet him.

One of the reasons his book was so popular and influential may have been because Oughtred loved to experiment with mathematical symbols, and regularly used signs to stand for quantities and

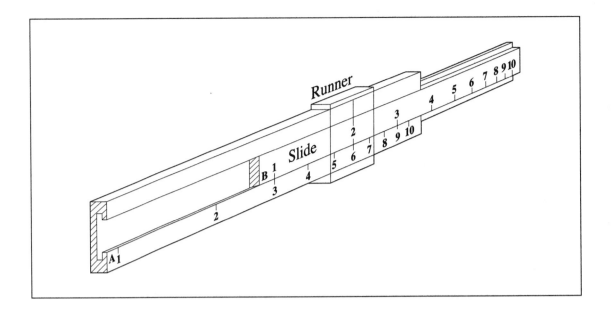

An illustration of a linear
slide rule.
*Illustration by Electronic Illustrators
Group. Reproduced by permission
of the Gale Group.*

operations. One symbol that caught on and has lasted until modern day was his use of the notation *"x"* for multiplication. Another was his use of four dots, "::," to symbolize proportion. Despite Oughtred's creation of what has been described as "a maze of symbols" only the equal sign and proportion symbol are still used today. For the most part, those that did not catch on were too complicated or unclear. Oughtred also used today's symbol for pi, π, in his book, but it did not stand for the **ratio** (see entry in volume 2) of the circumference to the diameter. Instead, it only stood for the circumference of a **circle** (see entry in volume 1).

Invents slide rule

Oughtred is also generally considered to be the inventor of both the circular and the linear slide rule. In its simplest form, a slide rule is a ruler with sliding scales that is used for rapid calculation since it reduces complex operations to simple **addition** (see entry in volume 1) and **subtraction** (see entry in volume 2). Although there is some disagreement as to whether he or his pupil, English mathematician Richard Delamain (1600–1644), was the first to invent the circular slide rule, there is no dispute concerning Oughtred's claim to the linear or straight slide rule, which he wrote about as early as 1621. Until the invention of the hand-held calculator in the late twentieth century, a slide rule was an invaluable calculating tool that hung from the belts of many engineers.

During the rest of his long life, Oughtred published six more books on mathematics, despite the fact that he was a minister and not a professional mathematician. Although his invention of the slide rule is considered his greatest accomplishment, his contributions toward an efficient form of mathematical symbols are also very important since such systems of notation were essential for mathematics making the transition from ancient to modern ideas.

William Oughtred

For More Information

Burton, David M. *The History of Mathematics: An Introduction.* New York: McGraw-Hill, 1999, pp. 321, 486.

O'Connor, J. J., and E. F. Robertson. "William Oughtred." *The MacTutor History of Mathematics Archive.* School of Mathematics and Statistics, University of St. Andrews, Scotland. http://www.groups.dcs.st-andrews.ac.uk/~history/Mathematicians/Oughtred.html (accessed July 23, 2002).

Scott, J. F. "William Oughtred." In *Biographical Dictionary of Mathematicians.* New York: Charles Scribner's Sons, 1991.

Born July 4, 1945
Denver, Colorado

American mathematician, educator, and writer

John Allen Paulos

John Allen Paulos is that rare type of mathematician who is as comfortable with words as he is with numbers. As a professor of mathematics whose goal is to help the general public understand numbers better and to lose its fear of math, he has become a public speaker, monthly columnist, and best-selling author. The success of his several books proves the value of his belief that math can be both fun and serious.

Once hated math

Although he was born in Denver, Colorado, John Allen Paulos (pronounced PAWL-ohs) grew up in Chicago, Illinois, and later Milwaukee, Wisconsin. Although Paulos was good in arithmetic, this was not due to the help or influence of his parents, Peter Paulos and Helen Sanavas, who he remembers as being indifferent to his mathematical abilities. However, his grandfather was interested in math, and together they would do all sorts of numerical tricks. This kind of number play was very different from the mathematics being taught at school.

As a youngster, Paulos did not enjoy going to his mathematics classes, even though he did well in arithmetic and loved to analyze and study baseball statistics. He remembers studying the newspaper's sports section each day, as well as closely reading a column of scientific questions and answers. It was not until middle school, however, that Paulos realized he really did not hate math at all, but instead strongly disliked the way it was being taught. School math seemed pointless to him, and as a grown-up, Paulos remembers that math was taught in a manner he described as "rote and mechanical." However, once he discovered that mathematics had a fun and a practical side—like figuring out a baseball pitcher's earned run average—he started to enjoy it a bit more.

As a middle schooler, Paulos must certainly not have been a typical student, since he read an essay by English philosopher and mathematician Bertrand Russell (1872–1970), which both impressed and inspired him with its logical thinking. Paulos soon found that once he came to appreciate both the usefulness of mathematics and its logical and more thoughtful side, he began to enjoy the subject more and more. Paulos continued to read more of Russell, who became an idol of his.

Mathematics as a career

Despite his new attitude toward mathematics, Paulos did not enter the University of Wisconsin as a math major. He took classes English, philosophy, and classical literature before finally settling on mathematics. He later realized that having a college background which he described as being "all over the place" eventually helped him write for a popular audience. After finally deciding on a career in mathematics, Paulos remained at the University of Wisconsin and received his Ph.D. in mathematics from that school in 1974. That same year, he began teaching mathematics at Temple University in Philadelphia, Pennsylvania, and eventually became a professor of mathematics there. Paulos married Sheila Schwartz in 1972 while he was still in graduate school, and eventually became the father of two children, Leah and Daniel.

As a new college instructor, Paulos followed the traditional publishing route and produced several scholarly papers on such serious subjects as mathematical **logic** (see entry in volume 2), **probability** (see entry in volume 2), and the philosophy of science. In the classroom, he did not want to make his students suffer with math the way he had, so he worked at making his class a positive and

John Allen Paulos

John Allen Paulos

even enjoyable experience. He found that using humor as a teaching tool helped him remove much of the mystery and even fear of numbers that many students had. In class, he tried to show that doing mathematics was much more than "just a matter of plugging into formulas and doing rote computations." He said that teaching math like that was similar to teaching English by having students diagram sentences all the time. He was sure that no student would ever come to appreciate literature that way. So in teaching math, he not only tried to use humor in class, but he also attempted to get his students talking and thinking about it in the classroom. Since he believes that the essence of mathematics is thinking (as opposed to simply doing something automatically without knowing why), he often gives his students writing assignments. While some of them may get annoyed, saying "This is a math course, why do I have to write?" he feels that the better they can explain a mathematical concept in writing, the better they will really understand it.

Becomes best-selling author

Besides his use of humor in class, Paulos had long felt that there was something different and even special about the sense of humor possessed by most mathematicians. He knew that they tended to take things literally, which often leads to humorous situations. He also knew that they used various techniques, like reducing something to its absurd level, that sometimes made for an odd way of looking at the world. Recalling his motivation to write about math and humor, he later stated, "I was always interested in jokes, in telling them as well as understanding what about them made them funny. I had even done some stand-up [a comedy act], but failed miserably."

Paulos's interest in math and humor led him to begin collecting and writing mathematical jokes and riddles, which he combined with cartoons and other drawings to produce his first popular book, entitled *Mathematics and Humor.* This fifty-eight-page book was written in 1978 for students and teachers; two years later, a larger, new edition was published, this time with a larger audience in mind. It was a bit more philosophical and serious in its study of math and humor. The book was successful, so he wrote another book on the same subject called *I Think, Therefore I Laugh.* Following the positive reception of these two books, Paulos decided to write a book that would explore the larger issue of the importance of numbers in everyday life. He believed that a fear of numbers and what he called a kind of mathematical illiteracy or an ignorance of math can be just as serious as not being able to read.

Innumeracy

Paulos called this ignorance of math "innumeracy" (pronounced ih-NOOM-ur-uh-see) and he described it as a numbers version of illiteracy. He argued that this type of unfamiliarity with math and numbers can lead to negative experiences in people's lives, from not knowing how to interpret the statistics they hear on the news to being taken advantage of in dishonest stock deals. In his book, *Innumeracy: Mathematical Illiteracy and Its Consequences,* he showed how those who suffer from innumeracy also have problems understanding things like logic and probability, and how this leads to them not being able to make sense of certain everyday things. Paulos believed that people who have this mathematical problem might not be able to understand their accountant or doctor. Further, they might not have the ability to be skeptical about what they hear and read, and could believe things without knowing how to look at them critically.

Paulos's book, which was published in 1989, appeared at a time when the American public was having growing concerns about the educational system. His book not only explained the concepts behind basic mathematics and made even the more difficult ones easier to understand, it introduced the notion of innumeracy and how serious it was to this nation. Paulos wrote that in today's world, the ability to understand math should be as basic and important as the ability to read. He also stated that the nation's problems or anxiety with math was the result of both the way it teaches math to young people and the way it fosters wrong attitudes and misconceptions about how hard a subject it is.

His next book, *Beyond Innumeracy: Ruminations of a Numbers Man,* addressed this problem, and Paulos wrote about seventy mathematical subjects and concepts in a manner that made them understandable, if not actually easy. Since then, he has written two similar books, *A Mathematician Reads the Newspaper* and *Once Upon a Number.* His work has proven to have a broad appeal, and his books have been translated into eleven languages. Paulos's fun

John Allen Paulos

and funny approach to math has resulted in multiple newspaper articles and radio and television appearances. Paulos believes that his message is finally beginning to get across, as interest in and appreciation of mathematics appears to be increasing across the country. He was especially encouraged to see that movies such as *Pi, Good Will Hunting,* and *A Beautiful Mind* had mathematicians as their central characters. If indeed, people begin to better appreciate the importance of mathematics in their lives and seek to do something about it, Paulos will be largely responsible.

For More Information

American Men and Women of Science. 20th ed. New Providence, NJ: R. R. Bowker, 1998, p. 1116.

Chang, Kenneth. "Writings of a Mathematician." *ABC News: Science.* http://abcnews.go.com/sections/science/DailyNews/paulos981202.html (accessed July 16, 2001).

Contemporary Authors. Vol. 136. Detroit: Gale, pp. 309–13.

John Allen Paulos and the Sporadic Exponent. http://www.math.temple.edu/~paulos/ (accessed July 16, 2001).

Paulos, John Allen. *Beyond Numeracy: Ruminations of a Numbers Man.* New York: Alfred A. Knopf, 1991.

Paulos, John Allen. *I Think, Therefore I Laugh: An Alternative Approach to Philosophy.* 2d ed. New York: Columbia University Press, 2000.

Paulos, John Allen. *Innumeracy: Mathematical Illiteracy and Its Consequences.* New York: Hill and Wang, 1988. Reprint, 2001.

Paulos, John Allen. *A Mathematician Reads the Newspaper.* New York: Basic Books, 1995.

Paulos, John Allen. *Mathematics and Humor.* Chicago: University of Chicago Press, 1980.

Paulos, John Allen. *Once upon a Number: The Hidden Mathematical Logic of Stories.* New York: Basic Books, 1998.

Born 1510
Tenby, Wales

Died 1558
London, England

Welsh-born English mathematician and educator

Robert Recorde

R obert Recorde has been called the founder of the English school of mathematical writers. English schools used his works as textbooks for more than one hundred years. As the first mathematician to introduce **algebra** (see entry in volume 1) in England, he is best known for inventing the equal sign.

Training in mathematics and medicine

Robert Recorde was born in Wales, one of the four countries that make up the United Kingdom, located west of England and bordering the Irish Sea. He was the second son of Thomas Recorde and Rose Johns. Little is known of his early years and family situation. Recorde graduated from Oxford University in 1531, and in that same year joined All Souls College, an institution specializing in religion, law, and medicine. The fact that he attended the prominent Oxford University indicates that his parents were not poor. Some time later, he moved on to Cambridge University and received a medical degree from that school in 1545. He eventually became a physician to England's King Edward VI (1537–1553) and Queen Mary I (1516–1558). It is likely that he lectured in mathematics at Oxford and Cambridge.

Robert Recorde

Influential in England

Within England, Recorde became the most influential writer on mathematics of his time. This was due in part to his willingness to write his books in simple, clear English rather than in the language of scholars, which was Latin. But, although this meant that more people in English-speaking countries could read his work, it also meant that he would not be read very much in any other country in Europe, which is why he did not gain much of an international reputation.

Recorde is known to have been a skillful teacher, and all of his mathematical works were intended to guide the student step by step through the many levels of mathematics. He believed that it was important for the student to learn and master things in the proper order, and his mathematical works were written and published in the order in which he intended them to be studied: arithmetic, plane **geometry** (see entry in volume 1), practical geometry, astronomy, and theoretical arithmetic and algebra.

Mathematical books

Recorde's first book, *The Grounde of Artes,* written in 1540 and expanded and republished many times, was also his most popular. It was a commercially successful arithmetic book and discussed the use of the abacus (a manual computing device), operations with Arabic numerals, and **fractions** (see entry in volume 1), among other useful subjects that were of interest to the everyday business world. Most of his books, including this one, were written in the form of a conversation between a teacher and pupil, which allowed him to carefully guide his student through each phase of a subject. Another popular aspect of his books was his extra effort to find the English equivalents for Latin and Greek technical terms. Finally, since he always proceeded in an orderly and logical manner, he put off dealing with difficult concepts and questions until the student had mastered the fundamentals.

In 1551, Recorde published *The Pathway to Knowledge,* a translation and rearrangement of the first four of the thirteen books of *Elements,* by Greek geometer **Euclid of Alexandria** (c. 325 B.C.E.–c. 270 B.C.E.; see entry in volume 1). Since Recorde believed that it was difficult for the beginning student to understand both the topic being taught and the reason why something was true, he separated the constructions or "things to be done" from the theorems or "things to be proved." His next textbook, *The Gate of Knowledge,* dealt with measurement and quadrants (astronomical devices for measuring altitude). This work has been lost, however, although some think that it may never have been published.

In 1556, Recorde's *Castle of Knowledge* appeared and offered an introductory treatment of Ptolemaic (pronounced tahl–oh-MAY-ik) astronomy, which was based on the work of Greek astronomer, geometer, and geographer **Claudius Ptolemy** (c. 100–c. 170; see entry in volume 3). Significantly, it also contained a short but favorable reference to the new, Sun-centered astronomy of Polish astronomer Nicolas Copernicus (1473–1543). This theory of astronomy went against the belief of Ptolemy and most others that stated that Earth was at the center of the universe.

Recorde's *Whettstone of Witte,* published in 1557, contained the "second part of the arithmetic" he promised in *The Ground of Artes,* and also contained some elementary algebra. It is in this text that he first introduced the use of two parallel line segments, "=," as the equal sign. As noted in *The MacTutor History of Mathematics Archive,* Recorde believed the two lines made sense "bicause noe 2 thynges can be moare equalle." Although the "=" symbol did not become immediately popular, it eventually became accepted. His only other book was a medical text called *The Urinal of Physick,* which told how to judge a person's condition by looking at his or her urine. It also contained sensible nursing advice.

Government duties and troubles

In addition to his mathematical and medical activities, Recorde also served the Crown in several capacities. In 1549, he was appointed comptroller (someone who supervises the spending of money) of the Bristol mint. In this position, he made several enemies when he refused to misuse money intended for the king. Because of this, he was accused of treason by his political enemies and was confined for sixty days. This dispute would eventually have serious consequences. In 1551, Recorde served as surveyor of the mines and monies in Ireland. When the mines proved to be unprofitable, Recorde was blamed and he was dismissed in 1553. When he tried to get his position back by bringing charges against his old and powerful enemies, he was sued by them for libel (an action that wrongly injures a person's reputation), and eventually lost his case. He was unable to pay the large fines and was put in prison (probably in 1557) where he wrote his will and died in 1558.

Robert Recorde

Recorde, who had five daughters and four sons, was an active supporter of Protestant Reformation, a growing movement against the Catholic Church. Some historians suspect that he may have been guilty of something more serious than debt, perhaps related to his involvement with the unprofitable Irish mines, since he was allowed to die in prison. In any event, Recorde's academic work was valuable and his reputation as an excellent mathematician lives on.

For More Information

Easton, Joy B. "Robert Recorde." In *Dictionary of Scientific Biography.* New York: Charles Scribner's Sons, 1980, pp. 338–39.

O'Connor, J. J., and E. F. Robertson. "Robert Recorde." *The MacTutor History of Mathematics Archive.* School of Mathematics and Statistics, University of St. Andrews, Scotland. http://www.groups.dcs.st-andrews.ac.uk/~history/ Mathematicians/ Recorde.html (accessed July 16, 2002).

**Born May 18, 1872
Trelleck, Monmouthshire, Wales**

**Died February 2, 1970
Merioneth, Wales**

Welsh-born English logician and philosopher

Bertrand Russell

The author of over forty books in many different fields, Bertrand Russell is one of the twentieth century's most influential intellects. A philosopher who also won the Nobel Prize for Literature, he made ground-breaking contributions to the foundations of mathematics by arguing that mathematics and **logic** (see entry in volume 2) are identical.

Bertrand Russell.
Courtesy of the Library of Congress.

A privileged orphan

Bertrand Arthur William Russell was born in Trelleck, Monmouthshire, Wales, one of the four countries that make up the United Kingdom. Wales is located west of England and borders the Irish Sea. Russell was born into an aristocratic (the upper class that governs a country) family that had long played an important role in the political, social, and intellectual life of Great Britain. His grandfather, Lord John Russell (1792–1878), twice served Queen Victoria (1819–1901) as her prime minister, and his father, also John Russell, was known as viscount (pronounced VYE-count; an English social rank above a baron and below an earl) Amberley. Bertrand Russell's mother, Katherine Stanley, was the daughter of a baron.

Bertrand Russell

The Russell family had long been associated with the Whig side of English politics, meaning they were liberal and favored a strong parliament (the British equivalent of the American Congress) and a non-powerful king or queen. Russell's parents were intelligent and progressive, and thus supported social and political reform, but neither was in good health. His mother died of a bacterial disease when he was only two years old, and his father died less than two years later. Russell also had an older brother named Frank.

After their deaths, the grandparents went to court to obtain custody of the brothers because they disapproved of the people that the parents had selected to raise their children. They had little trouble winning this case since the chosen couple admitted that they were atheists (pronounced AY-thee-ists; those who do not believe in the existence of God), which were disapproved of at that time. After the grandfather died in 1878, the Russell boys were raised entirely by their grandmother. Since she had little confidence in boarding schools, she arranged to have young Russell taught at her home by a series of governesses and tutors. Russell would later write in his *Autobiography* that although his grandmother had a rigid kind of morality, he was greatly influenced by her fearlessness, public spirit, and her indifference to the opinion of the majority.

Discovers Euclid and mathematics

While still living with his grandmother, Russell was introduced to *Elements,* by Greek geometer **Euclid of Alexandria** (c. 325 B.C.E.–c. 270 B.C.E.; see entry in volume 1), by his brother. Russell later wrote about this moment, saying, as noted in *The MacTutor History of Mathematics Archive,* "This was one of the great events in my life, as dazzling as first love. I had not imagined that there was anything so delicious in the world." The very bright Russell had no trouble at all understanding and mastering Euclid's theorems (statements or formulas obtained from a more general principal), but he became disappointed and annoyed to discover that Euclid started out with axioms (self-evident principles) that were not proved but simply had to be accepted on trust. The brilliant eleven-year-old refused to accept these axioms unless his brother could give him good reasons, but he finally had to give in when his brother refused to continue unless he agreed to accept them. This doubting attitude would remain with Russell all his life.

Russell left his grandmother's home at the age of fifteen when he took a training course for the scholarship examination to enter Trinity College, Cambridge. During this time, Russell was so lonely and miserable that he even thought about suicide but decided against it since he wanted to learn more mathematics.

Russell won a scholarship and did extremely well when he entered Trinity College in 1890 at the age of eighteen to study mathematics and philosophy. Although he later admitted that he "derived no benefit from lectures," he graduated first in his class in 1893. During his years there, he was elected to a small, informal group of intellectuals who called themselves the Apostles. There, he made friendships and met English mathematician and philosopher Alfred North Whitehead (1861–1947), with whom he would later write his major mathematical work.

Marries, travels, and begins writing

At the age of seventeen, Russell fell in love with Alys Pearsall Smith, who came from a wealthy Quaker family in Philadelphia, Pennsylvania, and after graduation they became engaged. His grandmother did not approve, and she arranged for him to take a position at the British Embassy in Paris, France, hoping he would become interested in politics and forget Alys. Russell found this job boring, however, and married Alys in 1894.

The following year, Russell won a six-year fellowship from Trinity that allowed him to travel and study. During this time, he formed a plan to write a series of books on the philosophy of the sciences, beginning with mathematics. He also traveled to the United States and Germany and wrote *German Social Democracy*, the first of his numerous books and pamphlets. This book showed his natural ability to investigate a subject quickly and then present it in clear and convincing language.

Mathematics and logic

In 1903, Russell wrote his first major mathematical work, *The Principles of Mathematics*. This was a highly original, but extremely technical, work in which he offered his belief that the foundation (basic principles) of mathematics could be deduced (conclusion drawn from a general principle) from only a few logical ideas. This was an ambitious task to try to prove, since if it could be done, the

Bertrand Russell

axioms that mathematicians have to accept would no longer be needed. This, he argued, would make both logic (the science of right or proper reasoning) and mathematics part of a single system.

Bertrand Russell

Russell's ideas on logic and mathematics were very different from the traditional ideas, since logic was always considered to be a tool that mathematics used to establish that something is valid or true. This notion fit with the other description of logic, which says that it is the process of combining statements to arrive at conclusions. In founding what was called the logistic school, Russell argued that instead of logic being a tool of mathematics, mathematics is a branch of logic.

In planning *The Principles of Mathematics,* Russell decided to write two volumes, the first containing explanations of his claims about logic and the second containing mathematical proofs. Russell's fellow Apostle group member Whitehead had been working on similar problems and the two scholars decided to work together on the second volume after Russell had finished the first. They worked together for nearly a decade, often sharing the same house, sending each other drafts and revising each other's work. The result was a separate, two-thousand-page work that they published in three volumes. A selection of almost any page of this work shows that it is full of mathematical symbols and that it is an extremely difficult book to read. Although most now agree that this was a monumental work, and that Russell argued to the end that mathematics and logic were identical, his great three-volume work did not firmly establish the correctness of his argument. In fact, there are other schools of thought besides the logistic school, and all are concerned with the seemingly simple question of what is mathematics. Today, many think that Russell and Whitehead went too far when they said that it was only through this sort of mathematics and logic that any certain knowledge could be attained.

Other careers and interests

In 1913, when the last volume of *Principles of Mathematics* was published, Russell was only forty-one years old, which was still fairly young for a person who would live to be nearly one hundred. During World War I (1914–18), Russell was an outspoken critic of England's participation in the war, and he actively worked against the drafting of young men into the army and published

Alfred North Whitehead,
with whom Bertrand Russell
collaborated on the three-volume
work entitled *Principles
of Mathematics.*
*Reproduced by permission of
AP/Wide World Photos.*

several pacifist (antiwar) books. He soon lost his lectureship at Trinity College because of his antiwar actions and eventually was imprisoned for six months in 1918 for antiwar activities. Russell used the time in prison to write another book on mathematics, *Introduction to Mathematical Philosophy.*

During the 1920s, Russell traveled to the Soviet Union, wrote about politics, and went on speaking tours. In 1921, Russell divorced his wife Alys and married Dora Black. His son, John, was

Bertrand Russell (center) joins in the fun with children from the experimental Beacon Hill School. *Reproduced by permission of AP/Wide World Photos.*

born soon thereafter, followed by his daughter, Katherine Jane. In 1927, Russell, who had been writing on education for several years, and his wife founded the progressive Beacon Hill School. This educational experiment lasted for five years.

In 1935, he divorced Black and married Patricia Spence the following year. They would have a son, Conrad Sebastien Robert, in 1937. From 1938 to 1944, Russell lived in the United States, where he lectured. He finally returned to Trinity College in 1944, where, in 1945, he published the work that would win him the 1950 Nobel Prize for Literature, *A History of Western Philosophy.*

During the 1950s and 1960s, he became somewhat of an inspiration to a new generation of young people opposed to war, and by then he was protesting the spread of nuclear weapons. At the age of eighty, after he divorced Spence, he remarried yet again, this time to Edith Finch. Ever the activist, he established the Bertrand

Russell Peace Foundation in his ninth decade, an organization that protested nuclear weapons testing and criticized American involvement in the Vietnam War (1954–75). Russell remained an active, public figure until his death at the age of ninety-seven. A man of remarkable intelligence and ability, his mathematical, philosophical, and political actions and works were guided by his belief in reason and common sense.

Bertrand Russell

For More Information

Ayer, A. J. *Bertrand Russell*. New York: Viking Press, 1972. Reprint, Chicago: University of Chicago Press, 1988.

Barrette, Paul. *The Bertrand Russell Gallery*. http://www.humanities. mcmaster.ca/~bertrand/ (accessed July 17, 2002).

The Bertrand Russell Archives at McMaster University. http://www. mcmaster.ca/russdocs/russell.htm (accessed July 17, 2002).

Broadbent, T. A. A. "Bertrand Arthur William Russell." In *Biographical Dictionary of Mathematicians*. New York: Charles Scribner's Sons, 1991, pp. 2180–88.

"Earl (Bertrand Arthur William) Russell—Biography." *Nobel e-Museum*. http://www.nobel.se/literature/laureates/1950/ russell-bio.html (accessed July 16, 2002).

Franceschetti, Donald R., ed. *Biographical Encyclopedia of Mathematicians*. New York: Marshall Cavendish, 1999.

Mathematicians and Computer Wizards. Detroit: Macmillan Reference USA, 2001.

Monk, Ray. *Bertrand Russell: The Ghost of Madness, 1921–1970*. New York: Free Press, 2001.

Monk, Ray. *Bertrand Russell: The Spirit of Solitude*. New York: Free Press, 1996.

O'Connor, J. J., and E. F. Robertson. "Bertrand Arthur William Russell." *The MacTutor History of Mathematics Archive*. School of Mathematics and Statistics, University of St. Andrews, Scotland. http://www.groups.dcs.st-andrews.ac. uk/~history/Mathematicians/Russell.html (accessed July 16, 2002).

Bertrand Russell

Russell, Bertrand. *The Autobiography of Bertrand Russell.* Boston: Little, Brown, 1967.

Russell, Bertrand. *A History of Western Philosophy, and Its Connection with Political and Social Circumstances from the Earliest Times to the Present Day.* New York: Simon and Schuster, 1945.

Russell, Bertrand. *Principles of Mathematics.* New York: W. W. Norton, 1938.

Born June 8, 1858
Lincoln, England

Died November 10, 1931
Cambridge, England

English analytical geometer and educator

Charlotte Angas Scott

Charlotte Angas Scott was a pioneer for the advancement of women in the field of mathematics. As the first woman in England to obtain a doctorate in mathematics, she was also the first dean and head of the mathematics department at Bryn Mawr College in Pennsylvania. Scott also was a founding member of the American Mathematical Society (AMS).

Charlotte Angas Scott.
Public domain.

Breaking "the iron mould"

Charlotte Angas Scott was the second of seven children born to Caleb Scott and Eliza Exley. She was born and raised in Lincoln, England, which is in the eastern part of that country. Her father was a minister of the Congregational Church and was also the president of Lancashire Independent College. Both her father and grandfather were described as "nonconformist Christians" since they supported social reform and the education of women.

Scott's father and grandfather encouraged young Scott to "break the iron mould" and seek a university education, a new and bold idea at the time. Her father took measures to prepare her, and he arranged for her to have the highest quality home schooling available, beginning with mathematics tutors at the age of seven. The tutors gave young Scott her first formal learning experience of mathematics, but she had already been introduced to that subject at home. Often her father would play mathematical games with his children, such as, "Think of a number. Double it. Add six. Take half of the answer. Subtract the original number. Your answer is three."

With this support and home tutoring, Scott won a scholarship in 1876 to Hitchin College, the first college for women in England, which had been established in 1869. At the age of eighteen, she joined its freshman class, which had only eleven students, and decided to become a mathematician. Hitchin College soon became one of the many colleges that made up Cambridge University and its name changed to Girton College and moved to within three miles of that large university.

Although Girton had been created as a sort of women's division at Cambridge, the university did little to make its female students' lives easy. In fact, the women attended class in almost primitive conditions. First, they had to walk three miles to get to Cambridge. Aside from the long daily distance, this presented another problem, since women were not allowed to walk alone and had to be accompanied by a chaperon. When the women got to class, not every professor would allow them to sit in on the lecture; those that did required the females to sit in class behind a screen so as not to distract the male students. This prevented the ladies from seeing the chalkboard. This was the educational climate that Scott and her fellow females had to endure in order to obtain an education.

"Scott of Girton"

Despite these hardships and discrimination, Scott worked and studied hard, and by 1880 she received permission to take the undergraduate final examinations called the "Tripos." These exams were not automatically given to women students since, officially, women at Cambridge were still not allowed to receive degrees. Once she obtained permission to take the exams, she sat through fifty hours of oral questioning spread out over one week. The results of the exams would determine which students received honors with their degree. When she was done, she had placed eighth among all the Cambridge students. The university tried to keep this fact a secret, but word soon spread throughout the campus of her accomplishment. Scott was not allowed to be at the award ceremony nor was her name even allowed to be read. However, at the ceremony, the names of the best students were read in the order they finished, and when her name was left out as the eighth best student in Cambridge, all the young men at the ceremony threw their hats in the air and shouted, "Scott of Girton! Scott of Girton!" This was such an unusual disruption that newspapers soon reported it throughout England. People were circulating petitions demanding equality for women at Cambridge. Scott was later honored by her friends in a private ceremony, and the university changed its examination policy for women the next year.

Career in the United States

Despite this publicity and attention, Cambridge still did not award Scott a degree. She took a second set of examinations at the University of London (which had begun granting degrees to women in 1876), and was able to receive her bachelor of science degree in 1882. At London, she had befriended English algebraist and geometer **Arthur Cayley** (1821–1895; see entry in volume 3), who became her mentor and helped her decide to pursue a doctorate in mathematics. Cayley was a believer in educational reform and had supported her ever since the 1880 incident at Cambridge. With Cayley guiding her graduate research, Scott received her Ph.D. from the University of London in 1885, the first degree of its kind to be earned by a British female. Both of Scott's degrees were of the highest possible ranking.

Pioneering achievements

When Scott received her Ph.D., there was only one other woman professor of mathematics in all of Europe, Russian mathematician **Sofya Kovalevskaya** (1850–1891; see entry in volume 2), who was teaching in Sweden. This fact did not persuade Scott to choose another career, however, and she decided to become a research mathematician, which required her to obtain a college-level teaching position. Fortunately, Bryn Mawr College opened in Bryn Mawr, Pennsylvania, in 1885, the same year she received her doctorate. This institution would become the first women's college in the United States to offer graduate degrees. With a recommendation from Cayley, Scott applied for a position there and became one of seven professors on its founding faculty. Bryn Mawr selected Scott to become the first head of its mathematics department, which at the time, consisted only of herself. Soon, however, the college attracted more students and began to grow, and Scott successfully lobbied for a series of reforms to the school's admission policies and entrance procedures.

One of her major accomplishments during this time was the role she played in establishing standards for admission. She persuaded her colleagues at other American colleges to pool their resources and offer common exams for admission. In 1901, the College Entrance Examination Board was instituted with her help, and she served as its first chief mathematics examiner in 1902 and 1903. At Bryn Mawr, she carried a heavy teaching load, in addition to working with graduate students as a thesis adviser. She was well liked and regularly demonstrated a genuine concern for all of her students.

In 1891, Scott was one of the first women to join the New York Mathematical Society. This later evolved into the American Mathematical Society (AMS) in 1895, and Scott was the only female member of its first council. The AMS became the major professional organization for research mathematicians in the United States and Scott served as vice president for the 1905–6 term.

In 1899, she became coeditor of the *American Journal of Mathematics,* an important research publication, and served in that capacity even after her retirement from Bryn Mawr in 1924. This position allowed her to positively influence the direction of American mathematical research. Her own research centered on the relationship of **geometry** (see entry in volume 1) and **algebra** (see entry in volume 1), a field that was later called analytic geometry. Her 1894 textbook on this subject, *An Introductory Account of Certain Modern Ideas and Methods in Plane Analytical Geometry* is still available and used.

Personal life

Scott spent forty years in the United States, and during her long time away from England she was visited regularly by her father and younger brother. During spring and summer breaks, she would visit her sisters in England and also travel to Europe where she had an expanding circle of relatives and mathematician friends in several major European cities. Scott never married and all her life was bothered with hearing problems. As early as 1884, Girton College wrote to Bryn Mawr that she was hard of hearing. By 1918, Scott was totally deaf, yet she could still lecture perfectly, as she used a graduate student to help her answer questions. When Scott retired, she moved back to England, close to Girton and Cambridge, where she died in 1931.

For More Information

Chaplin, Stephanie. "Charlotte Angas Scott." *Biographies of Women Mathematicians Web Site.* Agnes Scott College. http://www.agnesscott.edu/lriddle/women/scott.htm (accessed July 16, 2002).

Franceschetti, Donald R., ed. *Biographical Encyclopedia of Mathematicians.* New York: Marshall Cavendish, 1999.

Kenschaft, Patricia Clark. "Charlotte Angas Scott." In *Notable Women in Mathematics: A Biographical Dictionary.* Westport, CT: Greenwood Press, 1998, pp. 219–24.

O'Connor, J. J., and E. F. Robertson. "Charlotte Angas Scott." *The MacTutor History of Mathematics Archive.* School of Mathematics and Statistics, University of St. Andrews, Scotland. http://www.groups.dcs.st-andrews.ac.uk/~history/Mathematicians/Scott.html (accessed July 16, 2002).

Charlotte Angas Scott

Born October 31, 1815
Ostenfelde, Westphalia (now Germany)

Died February 19, 1897
Berlin, Germany

German analyst and number theorist

Karl Weierstrass

Karl Weierstrass.
Etching by Hans Thoma.
Reproduced by permission of the
Corbis Corporation.

Recognized as one of the greatest teachers of advanced mathematics, Karl Weierstrass is best known for setting rigorous (very strict) standards for calculus. What has become known as "Weierstrassian rigor" is a description of the methods employed by a man who was so methodical and logical that he was considered to be the "conscience" of mathematics.

Unable to study mathematics

Karl Theodor Wilhelm Weierstrass (pronounced VYE-ur-shtraws) was the oldest of four children born to Wilhelm Weierstrass and his wife, Theodora Vonderforst. His father was a secretary to a low-level government official, but when he became a tax inspector, he had to move his family frequently, so young Karl attended several different schools. Despite this, he was always a good student. When Weierstrass was twelve years old, his mother died and his father remarried a year later. His father was a well-educated man but he was very autocratic (acting with absolute authority), and when he noticed that his fifteen-year-old son was very good in math and had taken a part-time job as a bookkeeper, he decided that it would be best for him to become an accountant. His father reasoned that if he were a trained accountant, he could always obtain a secure position as a civil servant or government worker, and have a comfortable lifestyle.

Young Weierstrass, however, had different ideas. By the time he was a teenager, he was already reading serious mathematical journals, and knew that he very much wanted to study the subject that he loved, which was mathematics and not accounting. However, he was unable to disobey his father, and in 1834 he entered the University of Bonn to study law, finance, and economics. Since he had no interest in these subjects, Weierstrass spent most of his time at college participating in extracurricular activities. Instead of attending lectures, he worked on improving his fencing (dueling with swords), and spent more time reading mathematics books, socializing, and drinking in clubs than he did in school. Not surprisingly, after four years of such behavior, Weierstrass finally returned home without a degree, having left the university without even taking his exams.

Returns to school

His father was angry and shamed by his son's unwillingness to learn, but he was finally persuaded by a family friend to allow his son to go to another school and at least obtain a teaching certificate. So in 1839, after his son promised the school authorities that he would cooperate, his father sent him to the Theological and Philosophical Academy of Münster where he could obtain a secondary school teaching degree. This was the best thing that could have happened to young Weierstrass, since it was there that he met German mathematician Christoph Gudermann (1798–1851).

Weierstrass attended Gudermann's first lecture, after which he found himself to be the only student left in class. Gudermann had so intimidated everyone else that they simply dropped out of the class. Weierstrass soon became Gudermann's prize pupil, and he would be profoundly influenced by his teaching and ideas. In 1841, when Weierstrass was about to take the examination for his teaching certificate, he asked his professor if he would give him a mathematical problem that might normally be presented to a doctoral candidate. Weierstrass then gave an elegant and very original answer, and Gudermann was so delighted and amazed at the results that, as noted in *The MacTutor History of Mathematics*

Karl Weierstrass

Karl Weierstrass

Archive, he declared it to be "of equal rank with the discoverers who were crowned with glory." Unfortunately, Weierstrass did not learn of his professor's high opinion of his work until several years later, although he did know that Gudermann had recommended that he be granted a university position. The Academy chose not to do this, however, and Weierstrass simply took his teaching certificate and obtained a job as a mathematics and physics teacher at a secondary school (similar to a high school) in Deutsch-Krone, in west Prussia. He then moved to Braunsberg in 1848, where he became a lecturer at another secondary school, the Collegium Hoseanum. There, he would spend the next six years teaching such subjects as mathematics, botany, geography, history, calligraphy (handwriting), and even gymnastics.

Bursts from obscurity

As noted in *The MacTutor History of Mathematics Archive,* Weierstrass would later describe these fourteen years of teaching as "unending dreariness and boredom." After teaching students all day, he would usually go to a bar at night and work on some mathematics problems. For all these years, he was effectively cut off from any mathematical community, having no fellow mathematicians to talk to, no mathematical libraries to visit, and no extra money to pay for postage so he could write to other mathematicians. He was basically alone and remained totally unnoticed. To make these years even more miserable, Weierstrass often suffered attacks of dizziness, which left him violently ill. This is not surprising given the heavy demands of his poor-paying teaching job and the stress of trying to apply himself to mathematics in every spare moment. Sometimes, he would work on a mathematical problem all night long, unaware that the entire night had passed.

Fortunately, Weierstrass's life changed in 1853 when he became aware of the high opinion that the now-deceased Gudermann had of his work. This offered him renewed hope and gave him confidence in himself. During the summer of that year, he spent his vacation in his father's home and used the time to write a paper on the mathematical functions named after Norwegian algebraist **Niels Abel** (1802–1829; see entry in volume 1). This time, Weierstrass chose to publish his paper in a prestigious research journal called *Crelle's Journal.* The paper was accepted and appeared in print in 1854.

Sofya Kovalevskaya, good friend and prized student of Karl Weierstrass.
Courtesy of the Library of Congress.

Karl Weierstrass

The publication of Weierstrass's spectacular article seemed to electrify the entire mathematical community. Immediately, everyone recognized it as a work of a superior mathematician, but amazingly, it came from a man who was nearly forty years old and who had spent the last fourteen years teaching children. This fact so startled other mathematicians that Weierstrass became an instant celebrity. Hailed as a genius, he was given an honorary doctorate from the University of Königsberg and offered positions at the University of Berlin and the Royal Polytechnic School, both of which he accepted. Weierstrass was eventually overwhelmed by this incredibly sudden change of fortune, and in 1861 he collapsed and suffered a nervous breakdown.

Karl Weierstrass

Mentor to Kovalevskaya

When Weierstrass recovered and returned to teaching at Berlin, he began to attract mathematical students from all around the world. Over the years, the number of his students who went on to become major mathematicians grew to impressive proportions, and this was a real tribute to his famous teaching skills. Besides being one of the all-time great lecturers, Weierstrass always made himself available to students, and often insisted on paying for them at the local tavern they all would frequent.

One of his students, Russian mathematician **Sofya Kovalevskaya** (1850–1891; see entry in volume 2), became one of his closest friends as well. The two first met in 1870. During this time, the University of Berlin would not allow a woman to officially attend his lectures, so Weierstrass taught her privately for four years. Through his efforts, Kovalevskaya finally received her doctorate *in absentia,* meaning she was not present, from the University of Göttingen in 1874. For years after, the two kept up a correspondence. However, when he learned of her sudden death in Sweden at the age of forty-one, he was devastated and sadly burned all her letters. Her death so crushed him that he spent the last years of his life in a wheelchair before he finally died of pneumonia in 1897.

Mathematical legacy

Weierstrass was one of the few mathematicians who attained a major reputation without publishing a great deal. While he would contribute mightily to the development of Abel's key ideas, he became most famous for his insistence on exactness in all mathematical work. Weierstrass demanded that mathematics be based on clear and correct proofs, and this was the main reason he published so little. He needed to be absolutely sure that his work was based on the soundest mathematical foundation. While much of his work would become essential to the development of mathematical physics, his most important legacy is that he set certain standards of precision and exactness for mathematicians that would help shape the future of mathematics. Weierstrass never married and he was buried in a Catholic cemetery alongside his two sisters, with whom he had lived for much of his adult life.

For More Information

Biermann, Kurt R. "Karl Theodor Wilhelm Weierstrass." In *Biographical Dictionary of Mathematicians.* New York: Charles Scribner's Sons, 1991, pp. 2549–54.

Franceschetti, Donald R. *Biographical Encyclopedia of Mathematicians.* New York: Marshall Cavendish, 1999.

Golba, Paul. "Karl Weierstrass (1815–1897)." *IRA: Interactive Real Analysis.* http://www.shu.edu/projects/reals/history/weierstr.html (accessed July 16, 2002).

Mathematicians and Computer Wizards. Detroit: Macmillan Reference USA, 2001.

O'Connor, J. J., and E. F. Robertson. "Karl Theodor Wilhelm Weierstrass." *The MacTutor History of Mathematics Archive.* School of Mathematics and Statistics, University of St. Andrews, Scotland. http://www.groups.dcs.st-andrews.ac.uk/~history/Mathematicians/Weierstrass.html (accessed July 16, 2002).

Karl Weierstrass

Selected Bibliography

General sources

Asimov, Isaac. *Realm of Numbers.* Boston: Houghton Mifflin, 1959.

Ball, W. W. Rouse. *A Short Account of the History of Mathematics.* New York: Dover Publications, 1960.

Bergamini, David. *Mathematics.* Alexandria, VA: Time-Life Books, 1980.

Borman, Jami Lynne. *Computer Dictionary for Kids—and Their Parents.* Hauppauge, NY: Barron's Educational Series, 1995.

Boyer, Carl B., and Uta C. Merzbach. *A History of Mathematics.* New York: John Wiley & Sons, 1989.

Bunt, Lucas N. H., et al. *The Historical Roots of Elementary Mathematics.* Englewood Cliffs, NJ: Prentice Hall, 1976.

Burton, David M. *Burton's History of Mathematics.* Dubuque, IA: Wm. C. Brown Publishers, 1995.

Cajori, F. *A History of Mathematics.* New York: Chelsea, 1985.

Selected Bibliography

Dictionary of Mathematics Terms. New York: Barron's Educational Series, Inc., 1987.

Duren, Peter, ed. *A Century of Mathematics in America.* 3 vols. Providence, RI: American Mathematical Society, 1989.

Eves, Howard. *An Introduction to the History of Mathematics.* Philadelphia: Saunders College Publishing, 1990.

Flegg, Graham. *Numbers: Their History and Meaning.* New York: Schocken Books, 1983.

Friedberg, Richard. *An Adventurer's Guide to Number Theory.* New York: Dover Publications, 1994.

Green, Gordon W., Jr. *Helping Your Child to Learn Math.* New York: Citadel Press, 1995.

Green, Judy, and Jeanne Laduke. *A Century of Mathematics in America.* Providence, RI: American Mathematical Society, 1989.

Groza, Vivian Shaw. *A Survey of Mathematics: Elementary Concepts and Their Historical Development.* New York: Holt, Rinehart and Winston, 1968.

Heath, T. L. *A History of Greek Mathematics.* New York: Dover Publications, 1981.

Heddens, James W. and William R. Speer. *Today's Mathematics: Concepts and Methods in Elementary School Mathematics.* Upper Saddle River, NJ: Merrill, 1997.

Hirschi, L. Edwin. *Building Mathematics Concepts in Grades Kindergarten Through Eight.* Scranton, PA: International Textbook Co., 1970.

Hoffman, Paul. *Archimedes' Revenge: The Joys and Perils of Mathematics.* New York: Ballantine, 1989.

Hogben, Lancelot T. *Mathematics in the Making.* London: Galahad Books, 1974.

Humez, Alexander, et al. *Zero to Lazy Eight: The Romance of Numbers.* New York: Simon & Schuster, 1993.

Immergut, Brita. *Arithmetic and Algebra—Again.* New York: McGraw-Hill, 1994.

Julius, Edward H. *Arithmetricks: 50 Easy Ways to Add, Subtract, Multiply, and Divide Without a Calculator.* New York: John Wiley and Sons, 1995.

Katz, Victor J. *A History of Mathematics: An Introduction.* New York: HarperCollins College Publishers, 1993.

Kline, Morris. *Mathematics for the Nonmathematician.* New York: Dover Publications, 1985.

Kline, Morris. *Mathematics in Western Culture.* New York: Oxford University Press, 1953.

Miles, Thomas J., and Douglas W. Nance. *Mathematics: One of the Liberal Arts.* Pacific Grove, CA: Brooks/Cole Publishing Co., 1997.

Miller, Charles D., et al. *Mathematical Ideas.* Reading, MA: Addison-Wesley, 1997.

Moffatt, Michael. *The Ages of Mathematics: The Origins.* Garden City, NY: Doubleday & Company, 1977.

Rogers, James T. *The Pantheon Story of Mathematics for Young People.* New York: Pantheon Books, 1966.

Slavin, Steve. *All the Math You'll Ever Need.* New York: John Wiley and Sons, 1989.

Smith, David Eugene. *Number Stories of Long Ago.* Detroit: Gale Research, 1973.

Smith, David Eugene, and Yoshio Mikami. *A History of Japanese Mathematics.* Chicago: The Open Court Publishing Company, 1914.

Smith, Karl J. *Mathematics: Its Power and Utility.* Pacific Grove, CA: Brooks/Cole, 1997.

Stillwell, John. *Mathematics and Its History.* New York: Springer-Verlag, 1989.

Temple, George. *100 Years of Mathematics*. New York: Springer-Verlag, 1981.

West, Beverly Henderson, et al. *The Prentice-Hall Encyclopedia of Mathematics*. Englewood Cliffs, NJ: Prentice-Hall, 1982.

Wheeler, Ruric E. *Modern Mathematics*. Pacific Grove, CA: Brooks/Cole Publishing, 1995.

Wheeler, Ruric E., and Ed R. Wheeler. *Modern Mathematics for Elementary School Teachers*. Pacific Grove, CA: Brooks/Cole Publishing, 1995.

Wulforst, Harry. *Breakthrough to the Computer Age*. New York: Charles Scribner's Sons, 1982.

General biographical sources

Abbott, David, ed. *The Biographical Dictionary of Scientists: Mathematicians*. New York: Peter Bedrick Books, 1986.

Albers, Donald J., and G. L. Alexanderson, eds. *Mathematical People: Profiles and Interviews*. Boston: Birkhauser, 1985.

Albers, Donald J., Gerald L. Alexanderson, and Constance Reid. *More Mathematical People*. New York: Harcourt, 1991.

Alec, Margaret. *Hypatia's Heritage: A History of Women in Science from Antiquity through the Nineteenth Century*. Boston: Beacon Press, 1986.

Asimov, Isaac. *Asimov's Biographical Encyclopedia of Science and Technology*. Garden City, NY: Doubleday & Company, 1982.

Bell, Eric T. *Men of Mathematics*. New York: Simon and Schuster, 1986.

Biographical Dictionary of Mathematicians. New York: Charles Scribner's Sons, 1991.

Cortada, James W. *Historical Dictionary of Data Processing: Biographies*. New York: Greenwood Press, 1987.

Daintith, John, et al. *Biographical Encyclopedia of Scientists*. London: Institute of Physics Publishing, 1994.

Dunham, W. *The Mathematical Universe: An Alphabetical Journey through the Great Proofs, Problems, and Personalities.* New York: John Wiley & Sons, 1994.

Elliott, Clark A. *Biographical Dictionary of American Science: The Seventeenth Through the Nineteenth Centuries.* Westport, CT: Greenwood Press, 1979.

Gillispie, Charles C., ed. *Dictionary of Scientific Biography.* New York: Charles Scribner's Sons, 1990.

Grinstein, Louise S., and Paul J. Campbell, eds. *Women of Mathematics: A Biobibliographic Sourcebook.* New York: Greenwood Press, 1987.

Haber, Louis. *Black Pioneers of Science and Invention.* New York: Harcourt, Brace & World, 1970.

Henderson, Harry. *Modern Mathematicians.* New York: Facts on File, 1996.

Hollingdale, Stuart. *Makers of Mathematics.* London: Penguin Books, 1989.

Hudson, Wade, and Valerie Wilson Wesley. *Afro-Bets Book of Black Heroes From A to Z: An Introduction to Important Black Achievers for Young Readers.* East Orange, NJ: Just Us Books, 1997.

It, Kiyosi, ed. *Encyclopedia Dictionary of Mathematics.* Cambridge, MA: MIT Press, 1987.

McGraw-Hill Modern Scientists and Engineers. New York: McGraw-Hill, 1980.

McMurray, Emily J., ed. *Notable Twentieth-Century Scientists.* Detroit: Gale, 1995.

Metcalf, Doris Hunter. *Portraits of Exceptional African American Scientists.* Carthage, IL: Good Apple, 1994.

Millar, David, Ian Millar, John Millar, and Margaret Millar. *The Cambridge Dictionary of Scientists.* Cambridge, England: Cambridge University Press, 1996.

Morgan, Bryan. *Men and Discoveries in Mathematics.* London: John Murray Publishers, 1972.

Selected Bibliography

Morrow, Charlene, and Teri Perl, eds. *Notable Women in Mathematics: A Biographical Dictionary.* Westport, CT: Greenwood Press, 1998.

Muir, Jane. *Of Men and Numbers: The Story of the Great Mathematicians.* New York: Dover Publications, 1996.

Ogilvie, Marilyn Bailey. *Women in Science: Antiquity through the Nineteenth Century.* Cambridge, MA: MIT Press, 1986.

Osen, Lynn M. *Women in Mathematics.* Cambridge, MA: The MIT Press, 1974.

Pappas, Theoni. *Mathematical Scandals.* San Carlos, CA: Wide World Publishing/Tetra, 1997.

Perl, Teri. *Math Equals: Biographies of Women Mathematicians.* Menlo Park, CA: Addison-Wesley Publishing Company, 1978.

Porter, Roy, ed. *The Biographical Dictionary of Scientists.* New York: Oxford University Press, 1994.

Potter, Joan, and Constance Claytor. *African Americans Who Were First: Illustrated with Photographs.* New York: Cobblehill Books, 1997.

Reimer, Luetta, and Wilbert Reimer. *Mathematicians Are People, Too: Stories from the Lives of Great Mathematicians.* Palo Alto, CA: Dale Seymour Publications, 1995.

Ritchie, David. *The Computer Pioneers.* New York: Simon and Schuster, 1986.

Shasha, Dennis E. *Out of Their Minds: The Lives and Discoveries of 15 Great Computer Scientists.* New York: Copernicus, 1998.

Simmons, George F. *Calculus Gems: Brief Lives and Memorable Mathematics.* New York: McGraw-Hill, 1992.

Slater, Robert. *Portraits in Silicon.* Cambridge, MA: The MIT Press, 1989.

Spencer, Donald D. *Great Men and Women of Computing.* Ormond Beach, FL: Camelot, 1999.

Young, Robyn V., ed. *Notable Mathematicians: From Ancient Times to the Present.* Detroit: Gale Research, 1998.

Internet sites

Readers should be reminded that some Internet sources change frequently. All of the following web sites were accessible as of August 17, 2002, but some may have changed addresses or been removed since then.

The Abacus
http://www.ee.ryerson.ca:8080/~elf/abacus/

American Mathematical Society (AMS)
http://e-math.ams.org/

Ask Dr. Math
http://forum.swarthmore.edu/dr.math/

Athena: Earth and Space Science for K-12
http://inspire.ospi.wednet.edu:8001/

Biographies of Women Mathematicians
http://www.agnesscott.edu/lriddle/women/women.htm

Brain Teasers
http://www.eduplace.com/math/brain/

Canadian Mathematical Society
http://camel.cecm.sfu.ca/CMS

Eisenhower National Clearinghouse for Mathematics and Science
http://www.enc.org/

Explorer
http://explorer.scrtec.org/explorer/

Flashcards for Kids
http://www.edu4kids.com/math

Fraction Shapes
http://math.rice.edu/~lanius/Patterns/

Galaxy
http://galaxy.einet.net/galaxy/Science/Mathematics.html

MacTutor History of Mathematics Archive
http://www-history.mcs.st-andrews.ac.uk/history/

Math Forum: Elementary School Student Center
http://forum.swarthmore.edu/students/students.elementary.html

Math Forum: Math Magic!
http://forum.swarthmore.edu/mathmagic/

Selected Bibliography

Math Forum: Women and Mathematics
http://forum.swarthmore.edu/social/math.women.html

Math League Help Topics
http://www.mathleague.com/help/help.htm

Mathematical Association of America
http://www.maa.org/

Mathematical Programming Glossary
http://carbon.cudenver.edu/~hgreenbe/glossary/glossary.html

The Mathematics Archives
http://archives.math.utk.edu/

Mathematics Web Sites Around the World
http://www.math.psu.edu/MathLists/Contents.html

Mathematics WWW Virtual Library
http://euclid.math.fsu.edu/Science/math.html

Measure for Measure
http://www.wolinskyweb.com/measure.htm

Mega Mathematics!
http://www.c3.lanl.gov/mega-math/

Past Notable Women of Computing and Mathematics
http://www.cs.yale.edu/~tap/past-women.html

PlaneMath
http://www.planemath.com

Women in Math Project
http://darkwing.uoregon.edu/~wmnmath/

The Young Mathematicians Network WWW Site
http://www.youngmath.org/

Organizations

American Mathematical Society
201 Charles St.
Providence, RI 02904-2294
Internet site: http://www.ams.org

American Statistical Association
1429 Duke Street
Alexandria, VA 22314-3415
Internet site: http://www.amstat.org

Association for Women in Mathematics
4114 Computer and Space Science Building
University of Maryland
College Park, MD 20742-2461
Internet site: http://www.awm-math.org

Association of Teachers of Mathematics
7 Shaftesbury Street
Derby DE23 8YB England
Internet site: http://www.atm.org.uk/

Institute of Mathematical Statistics
P.O. Box 22718
Beachwood, OH 44122
Internet site: http://www.imstat.org/

Math/Science Interchange
c/o Department of Mathematics
Loyola Marymount University
7900 Loyola Blvd.
Los Angeles, CA 90045

Math/Science Network
Mills College
5000 MacArthur Boulevard
Oakland, CA 94613-1301
Internet site: http://www.expandingyourhorizons.org/

Mathematical Association of America
1529 18th Street N.W.
Washington, DC 20036-1385
Internet site: http://www.maa.org

National Council of Supervisors of Mathematics
P.O. Box 10667
Golden, CO 80401
Internet site: http://forum.swarthmore.edu/ncsm

Selected Bibliography

National Council of Teachers of Mathematics
1906 Association Drive
Reston, VA 20191-1502
Internet site: http://www.nctm.org

School Science and Mathematics Association
400 East 2nd Street
Bloomsburg, PA 17815
Internet site: http://www.ssma.org

Society for Industrial and Applied Mathematics
3600 University City Science Center
Philadelphia, PA 19104-2688
Internet site: http://www.siam.org/nnindex.htm

Women and Mathematics Education
c/o Dorothy Buerk
Mathematics and Computer Science Department
Ithaca College
Ithaca, NY 14850-7284
Internet site: http://www.wme-usa.org/

Index

A

Abacus, *1:* 81, 82, 83 (ill.)

Abel, Niels, *1:* 1 (ill.), **1–5**; *3:* 96; *4:* 160

Abel's theorem, *1:* 4

"The Absolute True Science of Space" (Bolyai), *3:* 20

Absolute zero, *3:* 160

Abstract, *2:* 410

Abstraction, *2:* 282

Académie Parisienne, *3:* 132; *4:* 61

Académie Royale des Sciences, *1:* 67, 68, 156, 222

Achenwall, Gottfried, *2:* 398

Acoustics, *1:* 189

Acton, Forman, *3:* 100

Acute triangle, *2:* 422

Ada (software language), *3:* 117

Addends, *1:* 8

Addition, *1:* **7–9**; *4:* 43

Addition symbol (+), *1:* 8; *2:* 295

"Aeolipile," *4:* 102

Aerospace technology, *1:* 201

Agnesi, Maria, *1:* 11 (ill.), **11–15**

Ahmes, *1:* 18; *2:* 416; *3:* **1–5**

Ahmes (Rhind) papyrus, *3:* 1–4

Aiken, Howard, *1:* 82, 215; *3:* 59; *4:* 1 (ill.), **1–7**, 5 (ill.)

Air, compressed, *4:* 102

Akademie der Wissenschaften, Berlin, *2:* 264

Alberti, Leone, *4:* 69

Alembert, Jean le Rond, d', *2:* 255, 255 (ill.)

Alexandria, Egypt, *3:* 138; *4:* 100

Algebra, *1:* **17–20**, 59, 61, 144, 171; *2:* 241

    Agnesi, Maria, *1:* 11–15

    Babylonia, *1:* 17

    Bari, Ruth Aaronson, *3:* 7–10

    Baxter, Agnes Sime, *4:* 23–26

    Bhāskara II, *4:* 28–30

    Boole, George, *1:* 49–52

    Boolean algebra, *1:* 49

    Bürgi, Joost, *1:* 33–37

    Cardano, Girolamo, *1:* 59–64

    Cayley, Arthur, *3:* 38–41

    China, *1:* 17

This is a cumulative index of *Math and Mathematics,* volumes 1–4. *Italic* type indicates volume number; **boldface** indicates main entries and their page numbers; (ill.) indicates photos and illustrations.

Index

Chuquet, Nicolas, *4:* 41–44
Descartes, René, *1:* 99–105
Egypt, *1:* 17
Falconer, Etta Zuber, *4:* 74–78
Fields, John Charles, *4:* 25
Galois, Évariste, *1:* 171–75
Hamilton, William Rowan, *1:* 207–11
Harriot, Thomas, *3:* 83–88
Khayyám, Omar, *4:* 110–14
Khwārizmī, al-, *2:* 241–45
Lagrange, Joseph-Louis, *2:* 253–58
Nash, John, *4:* 125–30
Noether, Emmy, *2:* 319–24
noncommutative algebra, *1:* 210
Oresme, Nicole d', *3:* 125–30
Pacioli, Luca, *4:* 43
Recorde, Robert, *4:* 141–44
Turing, Alan, *2:* 427–32
Viète, François, *3:* 164–67
word origin, *3:* 5
Algebraic notation
Oughtred, William, *4:* 132–35
Viète, François, *3:* 164–67
ALGOL (ALGOrithmic Language), *3:* 103; *4:* 16
Algorithm, *1:* 8; *2:* 241, 244, 394, 410; *4:* 42
word origin, *3:* 5
Algorithmic Language (ALGOL), *3:* 103
al-Khwārizmī. *See* Khwārizmī, al-
Allen, Paul, *3:* 103
Almagest (The Greatest) (Ptolemy), *3:* 138
al-Majisti (Great Work) (Ptolemy), *3:* 138
ALS (amyotrophic lateral sclerosis; Lou Gehrig's disease), *4:* 88
American Journal of Mathematics, *4:* 156
American Mathematical Society (AMS), *4:* 156

Amyotrophic lateral sclerosis (ALS; Lou Gehrig's disease), *4:* 88
Analog clock, *2:* 418
Analog computer, *1:* 82
Analysis, *1:* 134
Cantor, Georg, *1:* 53–57
Chang, Sun-Yung Alice, *3:* 43–46
Granville, Evelyn Boyd, *1:* 201–5
Hadamard, Jacques-Salomon, *3:* 77–82
Lambert, Johann, *3:* 105–10
Analytic functions, *3:* 78
Analytic geometry (coordinate graphing), *1:* 87–90, 89 (ill.), 99, 141, 143; *4:* 54, 156
Oresme, Nicole d', *3:* 125–30
Scott, Charlotte Angas, *4:* 153–57
The Analytical Arts Applied to Solving Algebraic Equations (Harriot), *3:* 87
Analytical engine, *1:* 33, 82; *3:* 113
Analytical Society, *1:* 32
Angle, *2:* 422
Apollonius of Perga, *1:* 142, 226; *2:* 237; *4:* **8–11**
The Apostles, *4:* 147
Appell, Marguerite, *4:* 38
Appell, Paul, *4:* 38
Applied mathematics
acoustics, *1:* 192
elasticity, *1:* 192
geodesy, *1:* 181
Germain, Sophie, *1:* 189–93
Granville, Evelyn Boyd, *1:* 201–5
Lovelace, Ada, *3:* 111–17
Arab scholars, Greek mathematics, *3:* 5
Arabic numerals, *1:* 148; *2:* 243–44; *3:* 5
Arago, François, *1:* 135
Archimedean screw, *1:* 22

B

Index

BASIC (Beginners All-Purpose Symbolic Instruction Code), *3:* 99, 103

Batch processing (computing), *3:* 101–2

Baxter, Agnes Sime, *4:* **23–26**

Beacon Hill School, *4:* 150

The Beautiful (Bhāskara II), *4:* 29, 30

A Beautiful Mind (Nasar), *4:* 130

Beeckman, Isaac, *1:* 100

Beginners All-Purpose Symbolic Instruction Code (BASIC), *3:* 99, 103

Bell, Eric T., *3:* 145; *4:* 126

Berlin Academy of Sciences, *2:* 264

Bernoulli, Daniel, *1:* 132, 133, 133 (ill.)

Bernoulli family, *1:* 132

Bernoulli, Jakob, *1:* 132; *2:* 344, 347 (ill.); *4:* 55

Bernoulli, Johann, *1:* 132, 132 (ill.), 133; *4:* 53, 53 (ill.)

Bernoulli, Nicholas, *1:* 133

Berry, Clifford E., *3:* 64

Bertrand Russell Peace Foundation, *4:* 150–51

Bessel, Friedrich Wilhelm, *4:* 83

Bessel functions, *4:* 83

Beyond Innumeracy: Ruminations of a Numbers Man (Paulos), *4:* 139

Bezout, Etienne, *1:* 153

Bhāskara I, *4:* 28

Bhāskara II, *4:* **28–30**

Bījaganita (Seed Counting or Seed Arithmetic) (Bhāskara II), *4:* 29

BINAC (Binary Automatic Computer), *3:* 62

Binary, *1:* 82

Binary Automatic Computer (BINAC), *3:* 62

Binary mathematics, *1:* 51

Binary notation, *2:* 262

Binary operation, *1:* 108; *2:* 294

Bit (computer term), *2:* 391

Black holes, *4:* 88–90, 89 (ill.)

Blackwell, David, *1:* 45 (ill.), **45–48**

Blaschke, Wilhelm, *1:* 72

Blockade of La Rochelle (Thirty Years' War), *4:* 61

Bluestockings, *3:* 112

Boiling point of water, in thermometric scale, *3:* 157, 160

Bolyai, János, *2:* 274; *3:* **17–22**

Bonaparte, Napoléon. *See* Napoléon Bonaparte

Boole, George, *1:* 49 (ill.), **49–52;** *2:* 282, 390; *4:* 33

Boole, Mary Everest, *4:* **31–34**

Boolean algebra, *1:* 49; *2:* 390

Bordin Prize, *2:* 250

Borel, Émile, *4:* 36 (ill.), **36–40**

Bossut, Charles, *1:* 154

Boundary, *1:* 28

Boyle, Robert, *2:* 260

Brahe, Tycho, *2:* 237

Bramer, Benjamin, *3:* 37

Breteuil, Gabrielle-Émilie le Tonnelier de, *2:* 263, 263 (ill.)

A Brief History of Time: From the Big Bang to Black Holes (Hawking), *4:* 90–91

A Briefe and True Report of the New Found Land of Virginia (Harriot), *3:* 84, 86

Briggs, Henry, *2:* 278; *3:* **23–27**

Briggsian (Common) logarithms, *3:* 25

Brightness, *3:* 108

Brinkley, John, *1:* 208

British Association for the Advancement of Science, *1:* 35

Brouncker, William, *3:* 28 (ill.), **28–32**

Brown, Robert, *1:* 113

Browne, Marjorie Lee, *1:* 201, 203, 203 (ill.)

Brownian motion, *1:* 113; *2:* 447

Bryn Mawr College (Bryn Mawr, Pennsylvania), *4:* 155

Bug (computer term), *1:* 215

Bulksley of Baron Hill, *4:* 106

Bureau of Ordnance Computation Project, *1:* 215

Index

E

d'Oresme, Nicole. *See* Oresme, Nicole d'

Dreyfus, Alfred, *3:* 79–80, 80 (ill.)

Index

Cayley, Arthur, *3:* 38–41

Chang, Sun-Yung Alice, *3:* 43–46

Chern, Shiing-Shen, *1:* 71–75

coordinate geometry, *1:* 103

Cramer, Gabriel, *4:* 51–55

Desargues, Girard, *4:* 60–64

Descartes, René, *1:* 99–105

descriptive, *3:* 108, 119–20

differential, *1:* 71, 72; *3:* 44

Egypt, *1:* 183; *2:* 415; *3:* 4

Escher, M. C., *4:* 66–72

Euclid of Alexandria, *1:* 125

Euler, Leonhard, *1:* 131–36

fiber bundles, *1:* 71

fractal, *2:* 287, 290; *4:* 38

graphs, *1:* 144

Halley, Edmond, *4:* 82–83

Hero of Alexandria, *4:* 99–103

Hippocrates of Chios, *3:* 89–93

Hypatia of Alexandria, *1:* 227

imaginary geometry, *2:* 274

Jones, William, *4:* 105–9

Khayyám, Omar, *4:* 110–14

Lambert, Johann, *3:* 105–10

Lobachevsky, Nikolay, *2:* 271–75

Maclaurin, Colin, *4:* 115–20

Mandelbrot, Benoit B., *2:* 287–91

Monge, Gaspard, *3:* 118–23

n–dimensional, *3:* 40

non-Euclidean, *2:* 271, 274; *3:* 20–21, 107

Oresme, Nicole d', *3:* 125–30

Pascal, Blaise, *3:* 131–36

plane geometry, *1:* 128; *4:* 101

practical, *3:* 34; *4:* 101

projective, *3:* 119–20, 132, 152–53; *4:* 60, 62–64

Ptolemy, Claudius, *3:* 137–41

Pythagoras of Samos, *2:* 349–53

Riemann, Bernhard, *2:* 379–83

solid geometry, *1:* 128; *4:* 101

Steiner, Jakob, *3:* 150–53

Thales of Miletus, *2:* 413–16

Geophysics, scientific, *4:* 83

Gerard of Cremona, *3:* 5

Germain, Sophie, *1:* 189 (ill.), **189–93**

German Mathematical Society, *1:* 56

German Social Democracy (Russell), *4:* 147

Gibbon, Edward, *1:* 228

Girton College (Cambridge), *4:* 154

Gödel, Escher, Bach: An Eternal Golden Braid (Hofstadter), *4:* 70

Gödel, Kurt, *1:* 195 (ill.), **195–99,** 198 (ill.)

Gödel's proof, *1:* 196

Golden Rectangle, *2:* 362

Goldstein, Herbert, *3:* 170

Gordon, George (Lord Byron), *3:* 111, 116

Gordon, Paul, *2:* 320

Göttingen Mathematics Institute, *4:* 48

Gram, *3:* 121

Grand Prix, Académie Royale des Sciences, *1:* 4, 67, 192; *2:* 255

Grandfather clocks, *1:* 165

Granville, Evelyn Boyd, *1:* 201 (ill.), **201–5;** *4:* 75

Graph theory
  Bari, Ruth Aaronson, *3:* 7–10

Graphic art
  Escher, M. C., *4:* 66–72

Graphs, *1:* 87, 144

Graunt, John, *2:* 398

Gravitation, *4:* 81

Gravity, specific, *3:* 86

Gray, Mary, *3:* 71–75

"Great Geometer." *See* Apollonius of Perga

Great Internet Mersenne Prime Search (GIMPS), *2:* 341

Great Library (Alexandria, Egypt), *4:* 100

Great Pyramid at Giza, *2:* 357, 359, 359 (ill.)

Index

Greater than symbol (>), *3:* 87

The Greatest (Ptolemy), *3:* 138

Greek astronomy, *3:* 138–39

Greek mathematics, *3:* 5

Gregorian calendar, *4:* 111–12

Gregory XIII (pope), *4:* 112

Grid, *1:* 88, 184

The Grounde of Artes (Recorde), *4:* 142

Group theory, *1:* 1, 174
> Galois, Évariste, *1:* 171–75

Gudermann, Christoph, *4:* 159–60

Guggenheim Museum (New York City), *4:* 63

Guide to Geography (Ptolemy), *3:* 139, 141

H

Hadamard, Jacques-Salomon, *3:* 77 (ill.), 77–82

Hahnemann, Samuel, *4:* 32

Halley, Edmond, *2:* 254, 315, 398, 399, 399 (ill.); *4:* 79 (ill.), **79–84,** 107

Halley's comet, *4:* 80–82, 81 (ill.), 82 (ill.)

Hamilton, William Rowan, *1:* 207 (ill.), **207–11**

Hand (measurement), *2:* 267

Hardware, *1:* 82, 84

Hardy, Godfrey Harold, *2:* 367

"Harmonia Macrocosmica," *3:* 140 (ill.)

Harmonic analysis, *1:* 156

Harmonics, *2:* 351

Harriot, Thomas, *3:* 83–88

Harvard Mark I. *See* Mark computers

Harvard University, *4:* 4

Hawking radiation, *4:* 88–90

Hawking, Stephen, *4:* 86 (ill.), **86–93,** 92 (ill.)

Hay Award, *4:* 98

Hay, Louise, *4:* **94–98**

Heat measurement, *3:* 160

Heat transfer, *1:* 155

Heinrich, Peter, *1:* 38

Heisenberg, Werner Karl, *3:* 40

Heliocentric theory, *2:* 235, 237; *3:* 129

Heliotrope, *1:* 181

Henry III (king of France), *3:* 165

Henry IV (king of France), *3:* 165

Hero of Alexandria, *4:* 99 (ill.), **99–103**

Herodotus, *2:* 415

Hero's formula, *4:* 101

Herschel, John, *1:* 32

Hertz, Heinrich, *1:* 113

Hieratic script, *3:* 2

Hilbert, David, *2:* 321, 446; *3:* 146, 147 (ill.); *4:* 47

Hill, A. Ross, *4:* 24

Hindu-Arabic numerals, *1:* 148; *2:* 243–44; *3:* 5

Hipparchus, *3:* 138

Hippocrates of Chios, *3:* **89–93**

A History of Western Philosophy (Russell), *4:* 150

Hofstadter, Douglas R., *4:* 70

Hollerith, Herman, *4:* 3, 3 (ill.)

Holmboe, Bernt Michael, *1:* 2

Hoover, Erna Schneider, *3:* 69

Hopper, Grace, *1:* 213 (ill.), **213–18,** 216 (ill.)

Hours, *2:* 418

House of Wisdom, *2:* 242

Howard, Thomas (earl of Arundel), *4:* 133

"Human calculator," *4:* 56–58

Humor, *4:* 138

Huygens, Christiaan, *1:* 165, 219 (ill.), **219–23;** *2:* 260

Hydraulic press, *3:* 134

Hydraulic screw, *1:* 22

Hydrogen bomb, *3:* 61

Hydrostatics, *2:* 406

Hypatia of Alexandria, *1:* 225 (ill.), **225–29**

Hyperbolas, *2:* 352; *4:* 10

Hypotenuse, *2:* 352, 356, 394, 422, 423

Hypothesis astronomia nova (Mercator), *4:* 123

Index

Kauffman, Nikolaus. *See* Mercator, Nicolaus

Kelvin, Baron of Largs (William Thomson; Lord Kelvin), *3:* 160–62, 161 (ill.)

Kemeny, John George, *3:* 100, 101 (ill.), 102–3

Kentron, *1:* 78

Kepler, Johannes, *2:* 235 (ill.), **235–40,** 238 (ill.); *3:* 35–36; *4:* 123

The Keys to Mathematics (Oughtred), *4:* 133

Khayyám, Omar *4:* 110 (ill.), **110–14**

Khwārizmī, al-, *1:* 18, 96; *2:* 241–45

Kolmogorov, Andrei N., *2:* 344

Kovalevskaya, Sofya, *2:* 247 (ill.), **247–51;** *4:* 155, 161 (ill.)

Kovalevsky, Sonya. *See* Kovalevskaya, Sofya

Kōwa, Seki. *See* Seki Kōwa

Kremer, Gerhard. *See* Mercator, Gerardus

Kronecker, Leopold, *1:* 55

Kulik, Yakov, *2:* 340–41

Kurtz, Thomas E., *3:* 99 (ill.), **99–104**

Kyoto Prize in Basic Science, *2:* 392

L

La Roche, Estienne de, *4:* 42

La Rochelle blockade (Thirty Years' War), *4:* 61

Lagrange, Joseph-Louis, *1:* 66, 154, 191; *2:* 253 (ill.), **253–58**

Lambert, Johann, *2:* 331, 377; *3:* **105–10**

Laplace, Pierre-Simon de, *1:* 208; *2:* 278, 301, 344, 348 (ill.)

Latin (language), *4:* 52, 142

Latitude, *1:* 88, 89; *3:* 138–39, 140

Lavoisier, Antoine, *2:* 257

Laws of planetary motion, *2:* 235, 239

Laws of probability, *1:* 59

Learning theory Boole, Mary Everest, *4:* 31–34

Least common denominator, *2:* 340

Leçons sur la théorie des fonctions (Lessons on the Theory of Functions) (Borel), *4:* 38

Lectiones geometricae (Geometrical Lectures) (Barrow), *3:* 15

Lectiones mathematicae (Mathematical Lectures) (Barrow), *3:* 15

Lectures on the Logic of Arithmetic (Mary Everest Boole), *4:* 34

Leibniz, Gottfried, *1:* 82; *2:* 259 (ill.), **259–65,** 282, 311–18; *4:* 55, 119, 119 (ill.)

L'Enfant, Pierre-Charles, *1:* 39

Length, *2:* 268

Leonardo da Vinci, *1:* 60

Leonardo of Pisa. *See* Fibonacci, Leonardo Pisano

Less than symbol (<), *3:* 87

Lessons on the Theory of Functions (Borel), *4:* 38

Levy, Paul, *2:* 288

Light, *1:* 113, 223; *3:* 108

"Light years," *4:* 83

"The Lightning Calculator," *4:* 56

Like fractions, *1:* 160

Līlāvatī (The Beautiful) (Bhāskara II), *4:* 29, 30

Lincean Academy, *1:* 167

Lincoln, Abraham, *1:* 128

Lindemann, Carl Louis Ferdinand von, *3:* 93

Linear equation, *1:* 18, 20; *4:* 54, 112

Linear measurement, *2:* **267–70,** 269 (ill.)

Linear slide rule, *4:* 134, 134 (ill.)

Liouville, Joseph, *1:* 174

Literary mathematics, *1:* 118

Lobachevsky, Nikolay, *2:* 271 (ill.), **271–75;** *3:* 21

Logarithmic tables, *2:* 278, 302

Logarithmotechnia (Mercator), *4:* 123

Index

National Bureau of Standards (NBS), *3:* 62

National Medal of Science, *1:* 75; *2:* 392, 449

National Medal of Technology, *1:* 217

National Physical Laboratory (NPL), *2:* 430

National Pi Day, *4:* 106

Natural numbers, *2:* 340, 394, 441, 442

Natural selection, *2:* 344

Navigation, *3:* 84; *4:* 82, 107
Harriot, Thomas, *3:* 84

N-dimensional geometry, *3:* 40

Negative numbers, *1:* 209; *2:* 231, 232, 376

Nemytskii, Viktor Vladimirovich, *4:* 21

Neumann, Hannah, *4:* 97

Neumann, John von, *1:* 84; *2:* 305 (ill.), **305–10,** 308 (ill.), 391, 429; *3:* 61; *4:* 38–39, 39 (ill.), 127

New Foundations of the Theory of Elliptic Functions (Jacobi), *3:* 96

A New Introduction to the Mathematics (Synopsis Palmariorum Matheseos) (Jones), *4:* 107

New World, *3:* 86

Newton, Isaac, *1:* 50, 208, 223; *2:* 261 (ill.), 262, 263, 311 (ill.), **311–18,** 314 (ill.), 387, 438; *3:* 14 (ill.), 14–15; *4:* 10, 81, 84, 108 (ill.), 117 (ill.), 119
Jones, William, *4:* 108–9
Maclaurin, Colin, *4:* 116–19

Night, *2:* 418

Nobel, Alfred Bernhard, *4:* 25

Nobel Prize, *1:* 113; *4:* 25, 130

Noether, Emmy, *2:* 319 (ill.), **319–24,** 323 (ill.)

Noncommutative algebra, *1:* 210

Non-Cooperative Games (Nash), *4:* 128

Non-Euclidean geometry, *2:* 271, 274; *3:* 107

Bolyai, János, *3:* 20–21

Northrop Aircraft Company, *3:* 62

Notation, *3:* 164–67; *4:* 133–34

Nuclear arms race, *4:* 128

Nuclear physics, *4:* 48

Number theory, *1:* 1, 128, 134, 144; *2:* 442, 443
Borel, Émile, *4:* 36–40
Cauchy, Augustin–Louis, *1:* 65–69
Dedekind, Richard, *3:* 51–55
Erdös, Paul, *1:* 119–23
Euler, Leonhard, *1:* 131–36
Fermat, Pierre de, *1:* 141–45
Fibonacci, Leonardo Pisano, *1:* 147–51
Lagrange, Joseph-Louis, *2:* 253–58
Ramanujan, Srinivasa I., *2:* 365–69
Robinson, Julia Bowman, *3:* 143–48
Weierstrass, Karl, *4:* 158–63
Wiles, Andrew J., *2:* 451–56

Numbers
composite, *2:* 340
even, *2:* 351
imaginary, *1:* 209
irrational, *1:* 128; *2:* 352, 375–77
mixed, *1:* 160
natural, *2:* 340, 441
negative, *1:* 209; *2:* 231
odd, *2:* 351
prime, *2:* 339–41
rational, *2:* 375–77
signed, *2:* 232
whole, *2:* 232, 441–44

Numerals, *2:* 244, 441–42

Numerator, *1:* 96, 160, 161; *2:* 372

Numerical scale, in thermometer, 156

0

Obtuse triangle, *2:* 422

Odd numbers, *2:* 351

Index

Olbers, Heinrich Wilhelm Matthaus, *4:* 83

On Painting and on Statuary (Alberti), *4:* 69

"On the Percussion of Bodies" (Maclaurin), *4:* 117

"On the Power of Gravity" (Maclaurin), *4:* 116

Once Upon a Number (Paulos), *4:* 139

Ophthalmoscope, *1:* 34

Oppenheimer, J. Robert, *2:* 308, 308 (ill.)

Optics, *2:* 240; *3:* 86
    Lambert, Johann, *3:* 108

Ore, Oystein, *1:* 214

Oresme, Nicole d', *3:* **125–30,** 129 (ill.)

"Organic architecture," *4:* 63

Orion Nebula, *1:* 221

Orthogonal projections, *3:* 120

Oughtred, William, *1:* 82; *2:* 296; *4:* 132 (ill.), **132–35**

P

Pacioli, Luca, *4:* 43, 43 (ill.)

Pangeometry, *2:* 274

Pappus of Alexandria, *1:* 142, 220; *4:* 9, 100

Papyrus, *1:* 18, 28; *2:* 422; *3:* 1
    Ahmes (Rhind), *3:* 1–4

Parabolas, *2:* 352; *4:* 10

Parallax, stellar, *4:* 83

Parallel postulate, *2:* 274; *3:* 19, 107; *4:* 113

Parallelogram, *1:* 28; *2:* 326, 362

Parallepipeds, *4:* 113

Parameters, *3:* 166

Paranoid schizophrenia, *4:* 128–29

Parchment, *2:* 340

Partitioning, *2:* 368

Pascal, Blaise, *1:* 82, 141, 144; *2:* 220, 261, 345, 345 (ill.); *3:* 131 (ill.), **131–36**

Pascal's principle, *3:* 134

The Pathway to Knowledge (Recorde), *4:* 142 ˙

Patronage, *4:* 106

Paulos, John Allen, *4:* **136–40**

Pearson, Karl, *2:* 398, 401 (ill.)

Pendulums, *1:* 165

Penrose, Roger, *4:* 89, 91 (ill.)

Pentagon, *2:* 326, 337 (ill.)

Pepys, Samuel, *3:* 30–31, 31 (ill.)

Percent, *2:* **371–74**

Percent symbol (%), *2:* 373

Percy, Henry, *3:* 85–86

Perfect tautochronism, *1:* 222

Perimeter, *1:* 28, 29, 78; *2:* 268, 269, **325–27,** 330, 334, 422, 434. *See also* Circumference

Perpendicular, *1:* 88; *2:* 422

Perspective, *1:* 184, 185; *4:* 62, 69

Pestalozzi, Johann Heinrich, *3:* 151 (ill.), 151–52

Philonides, *4:* 9

Philosophy
    Descartes, René, *1:* 99–105
    Hypatia of Alexandria, *1:* 225–29
    Leibniz, Gottfried, *2:* 259–65
    Pythagoras of Samos, *2:* 349–53
    Russell, Bertrand, *4:* 145–51
    Thales of Miletus, *2:* 413–16

Photoelectric effect, *1:* 113

Photometria (Photometry) (Lambert), *3:* 108

Physics, mathematical, *1:* 166, 189, 192
    12th century, *1:* 147
    17th century, *1:* 141
    18th century, *2:* 253
    Babylonia, *2:* 329, 356, 393
    Courant, Richard, *4:* 46–49
    discrete mathematics, *1:* 119
    Egypt, *1:* 160; *2:* 397
    Einstein, Albert, *1:* 111–18
    Euler, Leonhard, *1:* 135
    Galileo, *1:* 163–69
    Germain, Sophie, *1:* 189–93
    Greece, *1:* 127
    Hadamard, Jacques-Salomon, *3:* 81

Index

Ptolemaic system, *3:* 139, 140 (ill.)
Ptolemy, Claudius, *2:* 239; *3:* 137 (ill.), **137–41;** *4:* 10
"The Pulverizer," *1:* 33, 109
Punched cards, *1:* 33
Pure mathematics
    Cayley, Arthur, *3:* 40
    Cox, Elbert F., *3:* 47–50
Pythagoras of Samos, *2:* 349 (ill.), **349–53,** 350, 351 (ill.), 355, 357 (ill.), 372, 395; *3:* 90. *See also* Pythagorean theorem
Pythagorean theorem, *2:* 352, **355–59,** 358 (ill.), 395, 422. *See also* Pythagoras of Samos

Q

Quadratic equation, *1:* 61; *4:* 112
Quadrilateral, *2:* **361–63,** 363 (ill.)
Qualifier, *2:* 362
Quanta, *1:* 113
Quantity, *1:* 8, 17
Quantum mechanics, *1:* 174; *3:* 40; *4:* 48
Quantum physics, *2:* 307
Quantum theory, *1:* 113
Quaternions, *1:* 210
Queen Dido, *2:* 327
Quintic equation, *1:* 3
Quotient, *1:* 108, 109; *2:* 376

R

Radiation, Hawking, *4:* 88–90
Radical sign, *2:* 394; *4:* 43–44
Radicand, *2:* 394; *4:* 44
Radius, *1:* 78
Rahn, Heinrich, *1:* 109; *2:* 295
Raleigh, Sir Walter, *3:* 84–85, 85 (ill.)
Ramanujan, Srinivasa I., *2:* 365 (ill.), **365–69**
Random, *2:* 398
Rao, Ramachaudra, *2:* 367
Rao-Blackwell theorem, *1:* 47
Ratio, *1:* 78, 160; *2:* 326, 330, 344, 362, **371–74,** 376, 434
    Oresme, Nicole d', *3:* 128–29

Rational numbers, *2:* 375–77; *4:* 123. *See also* Irrational numbers
Rationes mathematicae (Mercator), *4:* 123
Recorde, Robert, *2:* 295; *3:* 87; *4:* **141–44**
Rectangle, *2:* 362
Refraction, *3:* 86
Regiomantanus (Müller, Johann), *4:* 113, 113 (ill.)
Regular polygon, *2:* 334
Relativity, *1:* 111
    general theory, *1:* 117; *2:* 319
    special theory, *1:* 114
Remainder (subtraction), *1:* 108; *2:* 410
Remington Rand Corporation, *3:* 64
Renaissance, *1:* 184
Rhind (Ahmes) papyrus, *1:* 17, 18; *3:* 1–4
Rhind, Alexander Henry, *3:* 1–2
Rhombus, *2:* 362
Ribet, Kenneth A., *2:* 455
Ricci, Ostillio, *1:* 165
Richelieu, Cardinal, *4:* 61
Richter scale, *2:* 278
Riemann, Bernhard, *2:* 275, 379 (ill.), **379–83**
Riemann surfaces, *2:* 381
Right angle, *2:* 356
Right triangle, *2:* 356, 422, 423 (ill.)
Rigor, *2:* 344
Robinson, Julia Bowman, *3:* 143 (ill.), **143–48;** *4:* 126
    Hilbert's "10th problem," *3:* 146
Rock and Roll Hall of Fame (Cleveland, OH), *1:* 186 (ill.)
Rod (measurement), *2:* 268
Roman Inquisition, *1:* 168
Roman numerals, *1:* 149
Roosevelt, Franklin D., *1:* 117
Royal Astronomical Society, *1:* 35
Royal Irish Academy, *1:* 52
Royal Society Computing Laboratory, *2:* 431

Transatlantic telegraph cable, *3:* 161–62

Transfinite numbers, *1:* 55

Translator (computer compiler), *4:* 15

Trapezoid, *1:* 28; *2:* 362

Treatise of Fluxions (Maclaurin), *4:* 118–20

Treatise on Demonstration of Problems of Algebra (Khayyám)

Triangle, *2:* 356, **421–26,** 423 (ill.), 425 (ill.); *4:* 101. *See also* Polygon; Pythagorean theorem

Triangulation, *2:* 422, 423

Trigonometria sphaericorum logarithmica (Mercator), *4:* 123

Trigonometry, *2:* 278, 299–303, 358, 421; *3:* 139

   Bari, Nina, *4:* 18–21

Triparty en la science des nombres (Three Parts in the Science of Numbers) (Chuquet), *4:* 42

Triple point of water, *3:* 162

Troy weight, *2:* 438

Turin Academy of Sciences, *2:* 254

Turing, Alan, *2:* 391, 427 (ill.), **427–32,** 428 (ill.)

Turing machine, *2:* 429

Two-term unit fraction, *3:* 4

U

Umar. *See* Khayyám, Omar

Unit fraction, *1:* 160; *3:* 4

Universal Automatic Computer (UNIVAC), *3:* 62

Up and Down (Escher), *4:* 71 (ill.)

U.S. Air Force Chapel (Colorado Springs, CO), *2:* 424 (ill.)

Utility, *4:* 54

V

Vacuum, *3:* 134

Vacuum tube, *3:* 60; *4:* 4

van der Hagen, F. W., *4:* 67

Variable, *1:* 103; *2:* 387; *3:* 166

Variations, calculus of, *2:* 253, 254

Veblen, Oswald, *1:* 74

Verne, Jules, *1:* 118

Vertex, *2:* 334, 422

Vesalius, Andreas, *1:* 61

Vibration patterns, *1:* 191

Victor Emmanuel III (king of Italy), *4:* 68

Viète, François, *1:* 142; *3:* 164 (ill.), **164–67**

Viviani, Vincenzo, *3:* 13

Volume, *1:* 25; *2:* 330, **433–36**

von Neumann, John. *See* Neumann, John von

Von Salis, Peter, *1:* 106

W

Wallis, John, *1:* 109, 110 (ill.); *3:* 30; *4:* 133

Ward, Seth, *4:* 133

Washington, D.C., *1:* 39

Water clocks, *2:* 418; *4:* 102

Water, freezing and boiling points, *1:* 58; *3:* 157

Water snail, *1:* 22

Watson, Thomas, Sr., *4:* 4

Weber, Wilhelm, *1:* 181; *2:* 381

Weierstrass, Karl, *2:* 249; *4:* 158 (ill.), **158–63**

"Weierstrassian rigor," *4:* 158

Weight, *2:* 435, **437–39**

Weyl, Hermann, *1:* 74

What Is Mathematics? (Courant), *4:* 49

Whettstone of Witte (Recorde), *4:* 143

Whirlpool galaxy M51, *4:* 89

Whitehead, Alfred North, *4:* 147, 149 (ill.)

Whole numbers, *2:* 232, **441–44**

Widmann, Johannes, *1:* 9; *2:* 295, 411

Wiener, Norbert, *2:* 445 (ill.), **445–50,** 448 (ill.)

Wiles, Andrew J., *1:* 144; *2:* 451 (ill.), **451–56,** 454 (ill.)

Wilhelm IV (duke of Hesse-Kassel), *3:* 34

Wilkins, J. Ernest, Jr., *3:* 168 (ill.), **168–71**

For reference

Not to be taken from the room.